£3-

Sundials

Standing universal ring dial by Thomas Heath, mid-eighteenth century.

Sundials

An Illustrated History of Portable Dials

Hester Higton

PHILIP WILSON PUBLISHERS

First published in 2001 by
Philip Wilson Publishers Ltd
7 Deane House
27 Greenwood Place
London
NW5 1LB

Distributed in the USA and Canada by
Antique Collectors' Club
91 Market Street Industrial Park
Wappingers' Falls
New York 12590

ISBN 0 85667 523 7

Designed by Peter Ling

Printed and bound in Italy by EBS, Verona

Opposite page:
*Gold pocket watch by Benjamin Lewis
Vulliamy, about 1847. The inside of the case
shows a number of different towns and cities
with time differences in GMT.*

CONTENTS

ACKNOWLEDGEMENTS

This book would never have seen the light of day without the support and hard work of the staff of the National Maritime Museum. I would particularly like to thank Alasdair MacLeod for the time he has devoted to its production. Thanks are also due to Jonathan Betts, Maria Blyzinsky, Tina Chambers, Gloria Clifton, Margarette Lincoln, Kristen Lippincott, Pete Robinson, Rob Petherick and Emily Winterburn. Above all, I am grateful for the opportunity provided by my time as the Sackler Research Fellow at the Museum to get to know the outstanding collection of sundials at the Royal Observatory and to work on this history.

A number of other people have helped me along my way in producing *Sundials* and my thanks go to Silke Ackermann, Jim Bennett, Giles Hudson, Stephen Johnston, Alison Morrison-Low, Anthony Turner, Gerard Turner, Tony Simcock, Liba Taub and Catriona West.

Lastly, I would like to thank my husband, Mike, for his assistance and encouragement along the way. This book is for him.

FOREWORD

Despite the fact that there is little concrete evidence about the origin of the sundial, we know that man has been using dialling instruments to mark the passage of time for thousands of years. This volume concentrates on portable dials, which have been known since ancient Egyptian times. In later centuries these dials became the kind of instrument that any respectable gentleman or gentlewoman might carry about the person in order to know the time of day. This book reveals a riveting story, combining science and art; commercial trade and piety.

Portable sundials form a major part of the instrumental holdings of the National Maritime Museum, and the collection is one of the most important in the world. Many of them came from the early munificent donations of Sir James Caird, who prized navigational and astronomical instruments as highly as maritime art. Numbered among the items are horizontal dials, ring dials, pillar dials, ivory and wooden diptych dials, magnetic and magnetic azimuth dials, Butterfield dials and so-called Regiomontanus dials, as well as astronomical compendia, quadrants, nocturnals, and one of the few surviving examples of a mediaeval *navicula*.

The riches of the Museum's collection are used to illustrate the tale told here. As such, it serves as a useful guide not only to the collection, but also to the complex and fascinating world of dialling itself.

Dr Kristen Lippincott
Director
Royal Observatory Greenwich

PHOTOGRAPHIC CREDITS

INTRODUCTION

Mention the word 'sundial' to most people and they will immediately think of an immovable object in the garden or on a wall. The Alex cartoon which shows three different sundials set up to provide the time in New York and Hong Kong as well as London may put the sundials to an unusual use, but the sundials themselves are exactly what we expect—and part of the humour comes from the fact that we all *know* that sundials are not accurate enough for Alex's requirements. Yet although the sundials provided by Peattie & Taylor will not carry out the job required of them, it would not have been difficult to set up three sundials that could have achieved this feat.

The idea of a portable sundial is probably even stranger. Surely it is only because a sundial is fixed in one place with the sun moving around it that it can tell the correct time. This whole relationship is destroyed when the sundial is picked up and moved to a different place. The idea of Fred Flintstone wearing a sundial on his wrist is quite clearly intended as a joke – no sundial could possibly work in this way, and certainly nothing that small could tell the time accurately from the sun. And yet, if we look at the

Cover illustration from Alex Plays the Game *by Charles Peattie and Russell Taylor.*

9

sundials which remain from previous centuries, we will find that the vast majority of them are portable and often pocket-sized. Many sundials were designed to be carried around and used to provide their owners with the time anywhere where the sun was shining. They were the pocket watches of the period before the eighteenth century, and were still used fairly regularly among rural communities until the beginning of the twentieth century. Not only that; until the nineteenth century they were often the most accurate timekeepers available; only the very rich could afford a reasonably accurate clock before 1800. But this aspect of the history of time-telling is very often overlooked, and few people are aware that it exists at all.

In this book I want to give you some idea of the wealth of portable sundials which exist, and the people who used them when they were first made. In the first chapter we will see how portable sundials have been in use for more than 3000 years (almost as long as people have kept time by any other means than a simple reference to the sun or moon) and we will look at some of the types available by 1000 AD. In the next chapter we will follow the development of more elaborate sundials, which were only available to the highest reaches of society. Many of these sundials were highly accurate, but also complex to use, requiring considerable skill on the part of the owner. Life was made somewhat easier with the introduction of portable versions of horizontal sundials (that is, sundials like the ones which appear in gardens and churchyards), and we will look at the changes brought about by the appearance of these new sundials in chapter three.

In chapter four we will see that the growth of affordable types of sundial allowed the creation of a whole industry for their production, particularly in Germany, but also in England and France. However, the real heyday of the portable sundial was the seventeenth and eighteenth centuries, when makers from England, France and the German states vied to produce the most popular forms. Chapters five and six turn our attention first to development in England and then to the changes made in France and Germany.

With the increase in the accuracy of watches, sundials fell into something of a decline, although they were still important for setting clocks and watches each day. Chapter seven traces the gradual eclipse of the portable sundial by mechanical timekeepers. However, in the final chapter we see how sundials are becoming increasingly popular again today, particularly as collectors' items, but also in the form of reproductions made to demonstrate how the people of past centuries regulated their days.

Chapter One

PILLARS AND RINGS

The sundial has long been described as the most ancient scientific instrument, and it is certainly the oldest timekeeper known to humankind. From earliest times people realised that the passage of the sun through the sky could be used to indicate the progress of the day from sunrise to sunset and that its position gave an approximate idea of the time of day. Similar information could be found from looking at the shadows which familiar objects such as a tree made on the ground. The length of the shadow and its direction would allow someone to work out whether it was morning or afternoon and how close it was to noon or to either end of the day. It must have been a very early development for sticks to be placed in the ground for the express purpose of showing time. Permanent marks could be made on the ground, indicating where the shadow of the stick lay at midday, sunrise and sunset and perhaps the mid-points in between.

However, it was probably not until the second millennium BC that the advantages of making such a timekeeper portable were realised. Poles in the

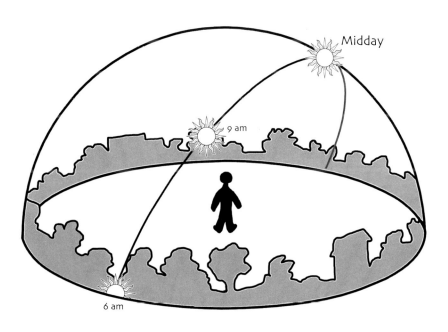

How the height of the sun is affected by the time of day.

ground were all very well, but you had to go back to them whenever you wanted to find out what time it was. A sundial which could be carried around meant that the time could be told wherever you were. So the portable sundial was born, and it was to endure as the most common form of timekeeper for the next 3,000 years. Of course, it is not clear why such accurate timekeepers were required at this time at all. Most measures of time were related to the rhythms of the human body or the duration taken to perform a particular task. The division of the daylight hours into artificial sections is not an obvious one to make, and we do not know exactly why this was first done.

Early sundials all depended on the height of the sun in the sky; the sun rises in the sky until it reaches the top of its arc at midday and then drops down towards the horizon. Correspondingly, the length of the shadow of a pole in the ground or any other form of marker alters through the day: the shadows of vertical markers decrease in length as the sun rises higher, while the shadow cast on a vertical surface by a horizontal marker increases towards midday. These changes are particularly marked in countries with low latitudes (those nearer to the equator), and sundials were first developed in the areas of the Middle East and North Africa.

The Sundials of Ancient Egypt

These early sundials are known as altitude dials (sundials are often referred to simply as dials) because they tell the time from the height of the sun in the sky. The earliest known surviving dial is of this type and it is also a portable dial. It was made in Egypt in the fifteenth century BC in the reign of Pharaoh Tuthmosis III, but little else is known about it. In all probability it belonged to a member of the priestly caste.

The dial consists of a stone rod, at one end of which is set a small vertical block. The lower piece is marked at unequal intervals along its length with a series of five small circles. These circles mark the hours of the day,

Two Egyptian altitude dials from the reign of Tuthmosis III, c.1500 BC.

although the indications are not for hours as we would think of them. There would once have been a plumb line hanging from the side of the vertical block, to make sure that the dial was placed in the right position to work properly. The dial is very simple to use: it is set on a flat, horizontal surface and turned until the end of the bar carrying the vertical block is pointing towards the sun. The shadow cast by the block now lies along the bar, and the time can be told from the position of the end of the shadow between the circles marked out on it.

This sort of sundial would have been reasonably accurate – it could have told the time to perhaps the nearest half hour, and certainly to the nearest hour. It might seem strange that such precision in timekeeping was required by early societies, where most tasks were simply continued until they were finished and the meal hours were sunrise, midday and sunset – all points which could be found without the use of a sundial. This is why it is likely that this sundial would have belonged to a high-ranking official or noble of the court who would have required an accurate timekeeper in order to fulfil meetings at specific times during the day. They would not have been able to determine the exact time by reading the position of the sun in the sky and would consequently have needed some form of mediating instrument.

Sundials in the Roman Empire

Similar uses were probably made of portable sundials in the Roman Empire. While there is a large gap in the history of these objects after the time of Tuthmosis III, a relatively large number of Roman portable dials remain. It seems likely that they would have belonged again to high-ranking officials or to wealthy nobles, people who either really did need to have a fairly accurate mark of the time, or people who acquired such objects as status symbols. All of the forms of sundial which survive from the Roman period are altitude dials, just as the Egyptian dials were. The reason that this sort of sundial was the only early form of sundial is that the construction of the lines to mark the hours is a great deal simpler on an altitude dial than on the sort of sundial which appears in gardens or on walls (we will return to this point at the beginning of chapter three).

Unlike the Egyptian dial, all the Roman portable dials which have been found were designed to be suspended from the hand rather than placed on a flat surface. They would be turned until the gnomon (the shadow-casting object, normally a small pin of metal) pointed towards the sun, at which point it would throw a shadow among the lines marked out on the sundial, so giving the time of the day. They are all small objects, never more than a few centimetres in length or diameter, designed to be carried in a pouch on the person so that they are available for use at any time.

The recent discovery of a Roman sundial has given us valuable new information about the types of dial which the Romans used. This dial was actually excavated in 1884, from the area around Este, a small town lying between Ferrara and Padua in Northern Italy. Around twenty ancient

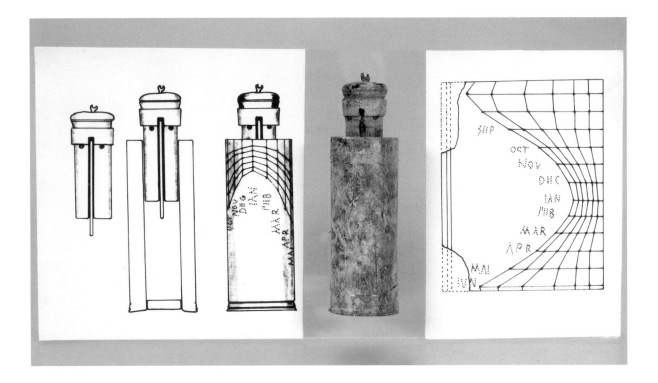

Roman cylinder dial, first century AD.

Venetian and Roman tombs were identified, including one described as the 'tomb of the physician'. The tomb was so called because among the artefacts discovered there were surgical and chemical instruments as well as ornaments, glass cups and an amber statue. One of the items unearthed was not identified by the original archaeologists; it was described by one of them in 1901 as a 'case' (astuccio). The object was placed on display in the Museum of Este, but it was not until 1984, exactly a century after the original discovery, that the 'case' was identified as a sundial.

This 'newly-discovered' sundial is vital to widening our knowledge of Roman dials, since it is the only known 'pillar' sundial from that period. Before it was identified, historians had thought that the pillar, or cylinder, sundial had been developed in the ninth or tenth century AD. However, the Este dial can be dated to the first century, showing that the form is far more ancient than was previously thought and that it may well have been developed by the Romans.

The dial consists of a short bone cylinder inscribed with lines and a cap (also of bone) which fits into the top of the cylinder. This cap carries two short, hinged bronze gnomons which can be stored within the dial when it is not in use. On top of the cap is a small bronze ring which would have been used for suspending the sundial when in use. Unfortunately, the gnomons are now so heavily corroded with oxidation products that they cannot be folded out from the cap to be used for indicating the time. However, it has been possible to use their measurements to calculate the latitude for which the instrument was designed.

The curving lines which are inscribed on the surface of the cylinder represent the hours of the day, just like the lines on an ordinary garden sundial. However, the hours shown by this sundial are different from modern timekeeping: they are seasonal hours. In early times, many civilisations divided daylight into twelve equal parts and the night similarly into twelve equal parts (though the day was the more commonly divided of the two). Since the length of daylight varies from season to season this meant that in summer each daylight hour was longer than in the winter. The hours were counted from the moment of sunrise, making it the beginning of the first hour, while noon was the sixth hour and sunset came at the end of the twelfth hour. Such systems were easier to use than one which depended on twenty-four equal hours in one day, because it was far harder to measure a period of sixty minutes than to measure out a twelfth of the daylight hours.

Each hour line on the cylinder represents two points in the day, except the line marking midday. This is because the sun is at the same altitude at the fifth hour as it is at the seventh hour and so casts the same shadow. The same is true for the fourth and eighth hours, the third and ninth hours and so forth. So with a dial of this sort you need to know whether it is morning or afternoon – not normally a problem, except when it is close to midday. The lines also curve and are crossed by vertical lines marked with the names of the months.

To understand the reason why the lines curve we must go back to thinking about the movement of the sun across the sky. In addition to the movement of the sun each day, we know that the sun is higher in the sky in summer than it is in winter: the midday sun of June is much higher than that of December and so it casts much shorter shadows. Thus a line on an altitude sundial which shows midday correctly in December will not show it correctly in June and we must make adjustments according to the date. The resulting line is a curve, and the appearance of these curves on the Este

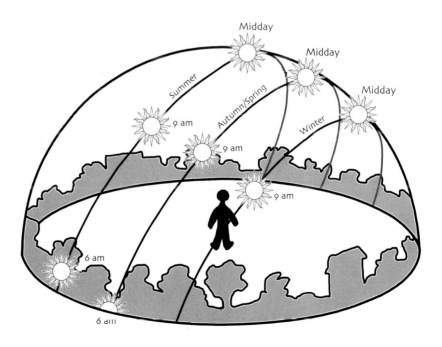

How the height of the sun is affected by the time of year.

dial means that it is adapted to cater for the seasonal changes.

The lines marked with the months allow the dial to be used correctly. The inscription on the cylinder is essentially a graph showing how the sun's altitude varies with the date, with a separate line being drawn for each hour of the day. The use of this instrument is straightforward: first the gnomon is hinged out so that it sits horizontally when the cap is replaced in the top of the cylinder. This dial is provided with two gnomons – the shorter one to mark the summer hours and the longer one to mark the winter hours – the correct gnomon must be chosen for the appropriate season. The gnomon is adjusted until it sits above the current month line and the dial is suspended from the ring in the top of the cap. Then the gnomon is turned towards the sun so that the shadow of the gnomon falls straight down the date line. The tip of the shadow indicates the time of day.

This dial has only twelve divisions for dates, making it rather difficult to align the gnomon with the correct point in the month. Interestingly, ten of these months are named but the seventh and eighth months are not. These are the ones which we now refer to as July and August and while it is possible that they were omitted due to lack of space it is equally possible that the omission was deliberate. July and August were freshly named in the second half of the first century BC, after Julius Caesar and Augustus Caesar respectively. So it is likely that the instrument dates from some time during the rule of Augustus. This would certainly fit with a date of c.50–100 AD, the date suggested for the tomb in which the dial was found.

The discovery of this pillar dial was exciting not only because it added to our knowledge of Roman forms of dial and because it showed that this form of dial was much older than had previously been thought, but also because the pillar dial is now clearly known as the sundial which has been longest in use as a timekeeper. Pillar or cylinder dials were probably produced regularly from the time of the Romans – they were popular in the Middle Ages, and although they were ousted from their position by the introduction of new forms of sundial during the Renaissance, they continued to be made from time to time, and seem to have retained a popularity among rural communities. They were still in use by the Basque shepherds of the Pyrenees at the beginning of the twentieth century, so they have a continuous history of use spanning two millennia.

As I have said, there were various other forms of sundial which first came into use during the Roman period, such as the ring dial which was made for many centuries. This was a simple ring of metal with a hole in one side and lines to mark the hours drawn out on the opposite edge of the inside of the band. These hour lines were drawn to accommodate the

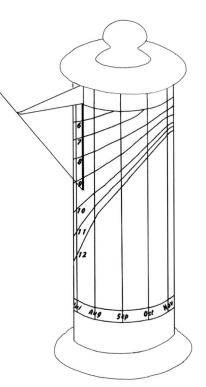

A cylinder dial in use.

changing seasons, just like those on the pillar dial, so that the ring dial could be used throughout the year. The 'gnomon' was provided by the hole in the side of the ring. However, with these sundials no shadow is cast on the hour lines. Instead, a narrow ray of light will pass through the hole and it is the spot of light, rather than a shadow, which marks the time from its position in the hour lines. The changing altitude of the sun through the year means that sometimes light will not fall through the hole if there is only one, or if it is fixed in position. To avoid this problem the ring may be equipped with two holes (one for summer months and one for winter) at different points on the band, and two sets of hour lines are marked out. More commonly, a central movable band carrying the hole was set around the ring, so that it could be rotated to set the dial for the date: a date scale of simple form was marked on the edges of the ring and the moveable band turned until the hole was at the correct date. Neither of these methods produced a particularly accurate dial and the rings were normally so small (never more than 3 or 4cm in diameter) that they usually could not tell the time to anything less than the nearest hour. However, they seem to have been popular and fairly cheap to make.

The ring dial and the pillar dial both continued in use for many centuries, though the pillar dial proved the most enduring. Their popularity is indicated by mentions in the literature of the Middle Ages and the Renaissance. A famous reference to the pillar dial can be found in Chaucer's *Canterbury Tales*: the 'Shipman's Tale' tells how a rich merchant is cheated by a monk, who first seduces the merchant's wife and then steals his money. The monk invites the wife to dinner with the words, 'And lat us dyne as

A plot of the hour lines for a cylinder dial. Notice how different the positions are for the same hour in winter and summer.

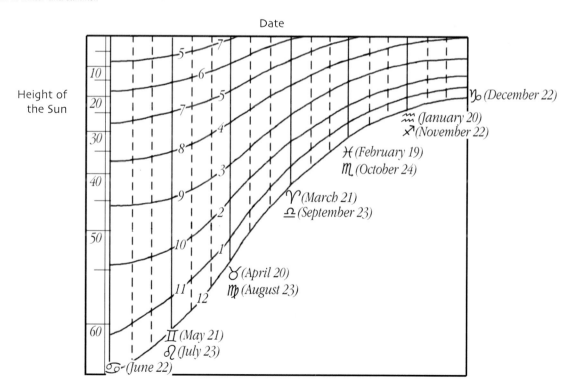

Date

Height of the Sun

♑ (December 22)

♒ (January 20)
♐ (November 22)

♓ (February 19)
♏ (October 24)

♈ (March 21)
♎ (September 23)

♉ (April 20)
♍ (August 23)

♊ (May 21)
♌ (July 23)

♋ (June 22)

17

Ring dial, unsigned, probably seventeenth century. The quality of engraving suggests that this dial was made by an amateur.

soone as that ye may; for by my chilyndre it is pryme of day'. The cylinder is, of course, a pillar sundial. References to sundials can also be found in Shakespeare: the most famous is that in *As You Like It*, act 2, scene vii, where Jaques gives an account of his meeting with the fool. During this encounter the fool

> '... *drew a dial from his poke,*
> *And looking on it with lack-lustre eye,*
> *Says very wisely "It is ten o'clock"; ...*'

It is not clear what sort of dial is referred to in this passage; the fact that ring dials were sometimes known as poke dials (that is, pocket dials) has led to the belief that it was probably a ring dial that was intended. However, most forms of portable sundial could be carried in the pocket, and so the description could fit any one of them. Nevertheless, both these quotations indicate the commonplace nature of sundials during the Mediaeval and Renaissance periods and the Chaucer reference shows the continuing popularity of the cylinder dial.

Sundials and the Mediaeval Church

Perhaps the largest group of sundial users during the first millennium AD were members of the church hierarchy. They would have used the common forms of dial, but we also have evidence of more elaborate creations. One unusual dial from this period was found in the grounds of Canterbury Cathedral in 1938. During the process of levelling the soil in the cloister garth, the workers unearthed a small pendant sundial, probably dating from the tenth century, and displaying evidence of Saxon workmanship. Saxon dials are relatively common as fixed objects in churchyards (such as the cross at Bewcastle). However, this is one of only two known examples of Saxon portable dials, and one of the few to have survived from the whole of the period between the Roman Empire and the Middle Ages.

The dial is made of silver and is roughly rectangular in shape. Inscribed on the sides are the words 'SALUS FACTORI' (salvation to my maker) and 'PAX POSSESSORI' (peace to my possessor). The upper end is capped by a gold mount for the pendant ring and chain (also made of gold). Amazingly, the gnomon was also found with the sundial – it is a small gold pin topped by an animal's head with green jewels for eyes, and a ball set between its jaws; a similar animal head is set at the end of the pendant chain. The gnomon can be stored in a hole in the centre of the sundial when it is not in use.

The scales on the sundial are extremely simple. Each side carries three columns which are marked for two months each (December and January, February and November, March and October on one side; April and September, May and August, June and July on the other). Each column also has a large hole at the top end and two further holes marked at points further down the column. The lower of these two points is used for marking midday, the upper for the end of the third and ninth hours of daylight (that is, mid-morning and mid-afternoon).

The Canterbury dial is used in much the same way as a pillar dial: the gnomon is pushed through the hole at the top of the relevant column so that the head touches the plate of the dial. The sundial is then hung from its pendant and turned so that the end of the gnomon points towards the sun; the shadow of the gnomon will now fall in the correct column for the date. The only times which can be read accurately from this dial are those marked by the holes for noon and the mid-morning/afternoon points; other times can be estimated but not accurately told. Indeed the layout of the Canterbury dial is not particularly accurate at all, and may well have been produced by very simple estimates of what the position of the shadow should be at particular hours.

A very similar sort of dial surviving from this period is kept at the Adler Planetarium in Chicago. It has similar markings, but is simpler in design and is made of bog-oak inlaid with bone. Unfortunately, the gnomon of this dial has not been found. It is worth noting that both dials concentrate on marking three particular points in the day, rather than showing all the hours. The reason for this is clear when we consider the patterns of worship which were standard in monasteries of this period (Christchurch,

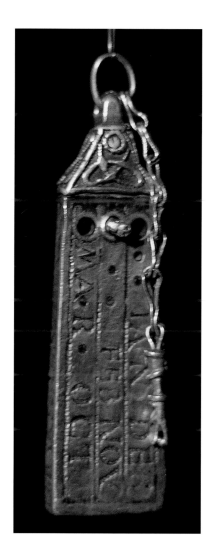

Saxon altitude dial, probably tenth century, found at Canterbury Cathedral.

19

Canterbury was, of course, a monastery as well as a cathedral until the Reformation). The monk's day was punctuated by eight points of prayer: matins, lauds, prime, terce, sext, none, vespers and compline. While the exact times of matins, lauds, prime, vespers and compline varied, they could generally be found from obvious points such as dawn (commonly lauds), sunrise (prime) and dusk (often the time of compline). However, terce, sext and none fell at the third, sixth and ninth hours of daylight – times which cannot easily be found without making use of some sort of time-telling device. The two Saxon dials can provide this information: when the shadow of the gnomon touches the upper hole it is time for terce or none (depending on whether it is morning or afternoon); when the shadow falls on the midday mark it is the hour of sext.

It is possible that dials like these were quite widely used in England for marking the times of the daylight services. The Chicago dial is probably more likely to have been the standard form, since it would have been relatively cheap to make. The dial that was found at Canterbury is altogether superior in composition, although the markings are rather less accurate than those on the Chicago dial. The use of precious metal and jewels for its construction indicate that it was made for a very wealthy owner. The inscriptions on the side further suggest that it may have been a gift from the maker to the owner.

It seems that most sundials made during the first millennium AD were intended for ecclesiastical use in some form or another. Scratch dials are occasionally found on Saxon church walls to mark the hours of prayer, and the portable dials also fit this pattern. Earlier in the chapter we saw dials which may well have been made for officials of the Roman Empire, but now the emphasis on timekeeping has shifted to the church. We will see in the next chapter that, as mechanical timekeepers were introduced (and were particularly used in monasteries from an early date), portable sundials began to become the possession of the secular individual, without reference to church or state, and eventually would become available to even the lower classes of society.

Chapter Two

HIGHER ALTITUDES

Introduction

While the simple altitude dials of the previous chapter continued to be made during the Middle Ages and bought by the merchant classes, a string of more elaborate designs began to appear. These were generally intended for sale to the nobility: they were often made of costly materials such as silver, gilt-brass or, occasionally, gold. Many of them were also quite complicated to use, often requiring some skill in mathematics, a subject which was generally only part of the education of aristocrats. So, while such possessions were prized for their expensive materials and the detailed ornamentation they also displayed the owner's knowledge of a complicated field of learning.

The types of dial which we will look at in this chapter are the *quadrans vetus*, the Regiomontanus dial, the *navicula* and the chalice dial. The chalice dial was a fairly standard form of altitude dial, but the other three were all a new form of sundial, developed from instruments used to measure the height of the sun, and adapted so that the height of the sun could be converted to show the time of day. All three instruments could also be used in different latitudes. This was an important step forward in the development of portable sundials, because it meant that they could be transported further North or South and still used to tell the time accurately.

The sun's height in the sky is affected by the latitude of the observor. The further one travels towards the equator the higher the sun in the sky, while the opposite is true when one moves towards the poles. Thus, a sundial which has been constructed for one particular latitude and depends on the patterns of the sun's movement in that latitude will not work if it is taken to another one. The first three dials which are described in this chapter, however, can be adjusted for different latitudes by processes which effectively move either the lines marking the hours or the point which is used to indicate the time. In so doing they compensate for the different paths which the sun makes in the sky at different latitudes.

The Quadrans Vetus

The name of this type of sundial simply means 'old quadrant', a term applied after the introduction of a different form of quadrant which became known as the 'new quadrant' and which we will meet in chapter five. The quadrant probably had its origins in Arabic countries, where examples have been dated to the eighth century; however, it did not become common in the West until after 1000 AD. It was the first type of dial which depended on a sighting of the sun in order for the time to be determined. In other words, the observer had to view the sun through the two sights set on the upper edge of the instrument to find the time. (More accurately, we should say that the sun's rays were allowed to fall through the upper sight so that they were seen shining on the hole in the lower sight: looking directly at the sun, even through a very small hole, could have blinded the user.) This type of arrangement also appears on Regiomontanus dials and *naviculae*, and sighting devices appear on various different types of dial in the seventeenth and eighteenth centuries.

The *quadrans vetus* consists of a quadrant of metal (normally brass) marked with curved hour lines and with two sights set along the right-hand edge of the instrument. Allowance is made for the date by means of a strip of metal which slides in an arc below the hour lines and is marked with the months of the year. The time indicator is a small bead set on the plumb line which hangs down from the quadrant's apex. This bead has to be adjusted for the correct date and latitude before the quadrant will show the correct

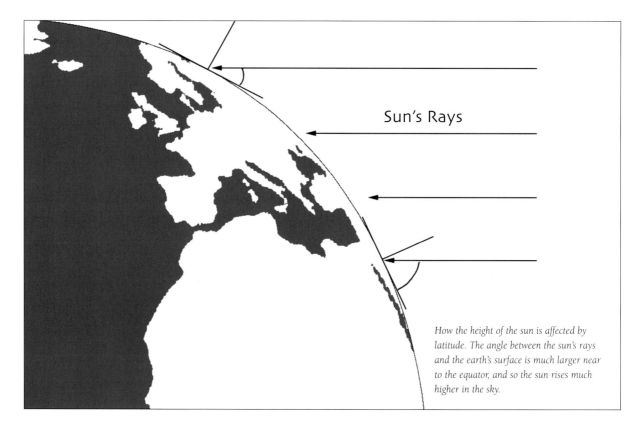

Sun's Rays

How the height of the sun is affected by latitude. The angle between the sun's rays and the earth's surface is much larger near to the equator, and so the sun rises much higher in the sky.

Quadrans vetus, English, fourteenth century.

time, and this is done with the help of the sliding date strip and the degree scale along the circumference of the quadrant, in a rather complicated process. First the centre point of the date scale must be set at the co-latitude of the place – this is equal to 90° minus the observer's latitude. So, if the observer's latitude is 52°, the centre of the date scale must be set to 38° on the degree scale. Once this has been done, the height of the midday sun is found from the point where the degree scale meets the current date. This value is needed for setting the bead on the string correctly – the string is laid across the quadrant so that it meets the degree scale at the point showing the height of the midday sun and finally the bead is moved along the taut string until it is positioned over the midday hour line.

After this rather longwinded setting-up process the quadrant is ready for use. The user raises it into a vertical position and adjusts the angle until the sun shines through the two sights on the right-hand edge. At the same time the hanging string must be clamped against the quadrant with one hand, so that when the user looks at the face of the quadrant, he will see the bead positioned among the hour lines at the correct time. This is quite a laborious method of finding the time, but the end result was usually more accurate than that given by pillar or ring dials.

Very few examples of the *quadrans vetus* survive, although there are many references to them in university texts of the Middle Ages. It is possible that only a few were ever made, since a form of quadrant often appears on the back of astrolabes from this period. However, the very good example illustrated here still shows the degree scale, date scale and hour lines very clearly, and even retains the two sights for viewing the sun.

The Regiomontanus Dial

While the Regiomontanus dial is a similar type of instrument to the *quadrans vetus*, it differs in two important ways. Firstly it was the first portable dial to be designed specifically to give the time in equal hours. The second difference, which is one of the reasons for the first, is that the Regiomontanus dial displays, in a two-dimensional form, the four-dimensional passage of the sun in space and time throughout the year. In some ways, this dial signifies an attempt to reduce the patterns of the heavens and their movements through the year into a flat diagram. Such reduction requires considerable knowledge of mathematics and mathematical astronomy, and was certainly not the province of those who made the simpler pillar and ring dials. It is not surprising, therefore, to discover that the Regiomontanus dial was designed by an astronomer.

A description of this dial first appeared in the 1474 almanac of the German astronomer Regiomontanus. This was the Latin name taken by Johannes Müller to demonstrate his connection with his home town of Könisberg. He is best known for his work in producing a summary of the greatest astronomical work of antiquity – Ptolemy's *Almagest* – this was a joint project with Georg Peurbach, and was not published until 1496, after the deaths of both astronomers.

While the use of twenty-four equal hours within a day had been standard among astronomers before the fifteenth century, it only came into use as an everyday system of timekeeping with the introduction of mechanical timekeepers, and even so it only slowly gained in popularity over the systems of unequal hours. Clocks were first developed at the very end of the thirteenth century and began to appear as prominent features in towns during the following two centuries. They were unable to mark unequal hours, because their mechanisms relied on an even rate of motion: so equal-hour timekeeping began to become a feature of mediaeval life.

The dial which Regiomontanus designed consisted of a rectangular plate with two sights set on the top edge. The plate itself was engraved with a set of parallel vertical hour lines which, like all other altitude dials, doubled for morning and afternoon hours. A scale marked according to date or the signs of the zodiac was laid out on the right-hand side of the plate and was used for setting the bead on the plumb line, just as we have seen for the *quadrans vetus*. This bead would show the time of day by its position in the hour lines just as that belonging to the quadrant did. However, there the similarity between the two instruments ends. The means for setting the

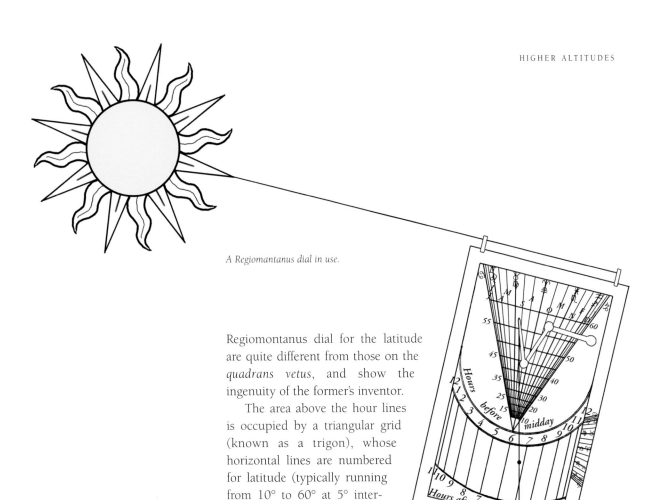

A Regiomantanus dial in use.

Regiomontanus dial for the latitude are quite different from those on the *quadrans vetus*, and show the ingenuity of the former's inventor.

The area above the hour lines is occupied by a triangular grid (known as a trigon), whose horizontal lines are numbered for latitude (typically running from 10° to 60° at 5° intervals, but this varies from one example to another) and whose 'vertical' lines mark the divisions of the zodiac. This scale was normally used rather than a date scale, since the position of the sun in the zodiac was seen as a more accurate measure than the date, which varied slightly with relation to the seasons because of the extra quarter day incorporated into every fourth year as the leap day of 29 February. The plumb line is fastened to an articulated arm which can move across the whole surface of the trigon and can therefore set the dial for any latitude and date which the observer requires (within the latitude constraints of the horizontal lines). The instrument is used by adjusting the articulated arm to the correct point, then adjusting the bead on the string to the appropriate date, and raising the instrument until the sun shines through the sights, at which point the bead will show the time by its position among the hour lines.

The Regiomontanus dial was the most accurate portable dial to be produced during the Middle Ages. However, the small number of surviving instruments suggest that it was not particularly popular, perhaps because the preparations for its use were complicated, or because it was difficult to make. Most of the examples which survive were made during the Renaissance and Early Modern periods, and seem to have been designed as showpieces for the rich. The picture here shows a fine example made by Johannes Krabbe in 1584 which was clearly designed for an aristocratic patron, probably a German judging by the large number of German towns which are included in the latitude list. Such latitude tables were an innovation with the arrival of dials which could be used in different latitudes. As the owner of the dial moved from town to town, he needed to have information about the latitude of the place where he was in order to be able to set his dial correctly. So we find Regiomontanus dials with lists of towns and cities throughout Europe and sometimes including other places as well: the Krabbe dial gives latitudes for Alexandria, Bethlehem and Constantinople, although Jerusalem (a common addition) is missing. It is unlikely that the owner would have travelled to such places, but they were places of historical and religious importance and so were included for their associations, rather than their usefulness.

The Navicula de Venetiis

Perhaps the most uncommon form of these aristocratic dials was the *navicula de Venetiis* which was produced during the late Middle Ages. The name means 'little ship of Venice' and comes from the shape of the dial, which is in the form of a mediaeval ship. The form seems to have appeared first in a fourteenth-century manuscript and it is similar to the Regiomontanus dial with straight lines supplying the hour marks and a bead on a string to mark the time. However, it was clearly developed independently from the Regiomontanus dial, since the early instruments date from before 1474 – the publication of Regiomontanus' design. The *navicula* seems to have been peculiar to England, despite its name, and the two surviving examples which appear with towns and latitudes marked on them provide information about places in England, and nowhere else.

The first mention in print of this unusual dial was in Oronce Finé's comprehensive work on sundials, *De Solaribus Horologiis et Quadrantibus*, published in Paris in 1560. This was one of the most important early works on sundials but its appearance came long after the time during which the *naviculae* had been made. All four mediaeval examples of the instrument date from the fifteenth and early sixteenth centuries. However, Finé's book may have rejuvenated interest in this form of dial, since two examples from the seventeenth century also survive. Another revival of interest came in the second half of the eighteenth century when an article appeared in the *Gentleman's Magazine* of 1787, giving details for construction. It is possible that *naviculae* were produced as a result of this article, but we have no record of such dials having been made.

The National Maritime Museum is fortunate to own one of the three *naviculae* which date from the fifteenth century. It was found by chance in July 1989 at Sibton Abbey, near Saxmundham in Suffolk, where it had been buried in the soil for five centuries. The remarkably good condition of the dial when it was retrieved suggests that it was probably lost shortly after it was made. All parts except for the plumb line and bead have survived, and the markings are still very clear. It is likely that it was owned by a rich

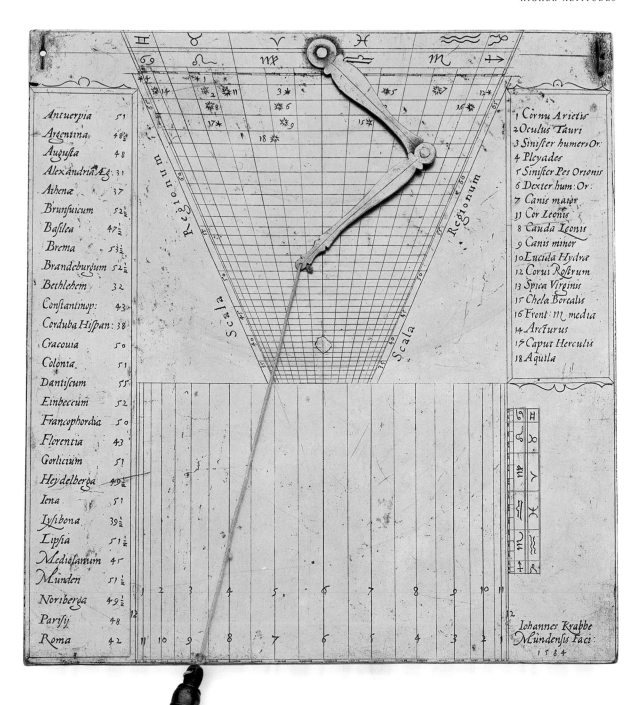

Antuerpia	51
Argentina	48½
Augusta	48
Alexandria Æg:	31
Athenæ	37
Brunfuicum	52½
Basilea	47½
Brema	53½
Brandeburgum	52½
Bethlehem	32
Constantinop:	43
Corduba Hispan:	38
Cracouia	50
Colonia	51
Dantiscum	55
Einbeccum	52
Francophordia	50
Florentia	43
Gorlicium	51
Heydelberga	49½
Iena	51
Iysibona	39½
Lipsia	51½
Mediolanum	45
Munden	51½
Noriberga	49½
Parisij	48
Roma	42

1 Cornu Arietis
2 Oculus Tauri
3 Sinister humers Or:
4 Pleyades
5 Sinister Pes Orionis
6 Dexter hum: Or:
7 Canis maior
11 Cor Leonis
8 Cauda Leonis
9 Canis minor
10 Lucida Hydræ
12 Corui Rostrum
13 Spica Virginis
15 Chela Borealis
16 Front: m. media
14 Arcturus
17 Caput Herculis
18 Aquila

Iohannes Krabbe
Mundensis Faci:
1584

Regiomontanus dial by Johannes Krabbe, 1584.

visitor to the abbey, though it could perhaps have belonged to the abbot himself. Like the Regiomontanus dial it could be used in a number of different latitudes, and so was clearly intended for an owner with reason to travel. It is appropriate that a travelling dial of this kind should be made in the form of a ship, a common form of transport between England and the lands of the Southern Mediterranean.

The Sibton Abbey *navicula* consists of a brass plate in the form of a ship whose hull is topped by a castellated poop and forecastle, and a separate mast. The castellations are not merely there for decoration (echoing the

Navicula, mid-fifteenth century, front, showing main sundial.

design of the ships after which the dial was named), but also carry the pin-hole sights which were used for lining the *navicula* up with the sun. The front of the dial is engraved with the hour lines – these are all vertical just like those on the Regiomontanus dial, which works on very similar principles. In common with that dial there is a scale on the side of the instrument which can be used for setting the bead, according to the signs of the zodiac.

The reverse of the instrument carries a second dial, in the form of a *quadrans vetus*, which could be used to provide the unequal hours of the day from sunrise to sunset, while the front of the dial gave the equal hours. The quadrant is divided into degrees along the lower edge, so allowing altitudes of the sun or of objects on the Earth to be measured in terms of degrees. A second quadrant on the left gives markings along the straight edges of the bottom and left-hand side. These are divided into twelve equal parts, and allow heights of objects to be taken in terms of the ratio between

Back of navicula with the quadrans vetus *and shadow square.*

their height and the distance from the observer. Such forms of measurement were important at a time when very few people had the ability to use trigonometry for converting angles to heights and distances. The problems of using multiplication and division for finding these values from a ratio were quite complicated enough without the involvement of trigonometry.

The function which was performed by the trigon on the Regiomontanus dial is here supplied by the mast of the ship. The mast is free to pivot around a point near the top of the main plate, which is split to allow the mast to pass through its centre. It is marked with a latitude scale up the centre which runs from 20° to 65° N, allowing use from the Sahara desert to central Scandinavia. The date scale is provided by a zodiac scale along the base of the ship, against which the mast can be set by rotating it around its pivot. This *navicula* provides assistance for using a zodiac scale: a table on the back gives the date at which the sun moves into each sign of the zodiac, permitting an easy conversion without referring to astronomical tables.

A collar on the mast holds the top end of the plumb line and it is moved up or down the mast so that it is against the correct latitude. Once this has been done the mast is set to the appropriate date, after which the bead is set against its own scale on the right hand side of the instrument. Then, as with the Regiomontanus dial, the *navicula* is raised until the sun shines through the holes in forecastle and poop, at which point the bead will indicate the time by its position on the hour lines.

The back of the mast of this *navicula* carries a list of five towns with their latitudes, a feature which appears on only one other example – that in the Musée d'Histoire des Sciences in Geneva. The towns are listed in Latin and include Eboracum (York), Northampton, Oxon (Oxford), Londinium (London) and Exon (Exeter). The latitudes which are given for the towns vary in accuracy: those for Northampton, Oxford and London are all within six minutes of a degree of the correct values, but those for Exeter and York are almost twenty minutes of a degree out. It suggests that the value for London had been calculated by astronomical means, but that the others were based on the distance of the towns North or South of London, and so the error increases with the distance away from the capital. The choice of towns for inclusion on the instrument seems unusual – London and York could perhaps be expected as standard places but Northampton in particular is a strange addition. It is probable that the original owner had links to these places and had requested that they were included on the *navicula* when it was made (unless he made it himself, in which case he could have selected the towns on the mast at will). They are once again an indication that this dial was no doubt the property of a rich man who could commission an instrument exactly to his requirements.

The Chalice Dial

Chalice dials were the most ornate form of sundial to appear during the Renaissance period, and were possibly made more for show than for use, although all the known examples would have been perfectly serviceable as time indicators. The hour lines are marked on the inside of a precious metal drinking cup and the gnomon is provided by the point of a vertical rod set in the centre of the chalice. A large number of chalice dials work on the standard principle of the projection of the gnomon's shadow through the air. However, they could also be designed for use when the cup was filled with water (or even white wine). Such dials as these would require a different set of lines to be marked, since the shadow of the gnomon is bent when it hits the surface of the liquid as a result of refraction. They would have been more complicated to make than the dials which were designed to be used empty because the mathematics involved in projecting the refracted lines is more complicated. It is possible that such dials were largely made by experiment, on a trial and error basis. Whatever their origin they demonstrate the skill of the maker, as well as providing an added novelty to their use.

The earliest surviving chalice dial was made in Aldersbach, a small town near Passau on the German-Austrian border, in 1554. It is signed in a cartouche within the bowl of the cup, below the hour lines:

'*BARTHOLOMEVS*
ABBAS ALDERS
PACENSIS
FACIEBAT
ANNO MDLIIII'.

Translation reveals that the object was made by Bartholomew, Abbot of Aldersbach. He further marked the base of the cup with the latitude of the dial: '*HOROLOGIVM IN CRATERE AD ELEVA PO 48*' (dial in a chalice for an elevation of the pole – i.e. latitude of 48°). The latitude of Aldersbach is, in fact, 48° 35′ so a dial for 48° would be sufficiently accurate for use in the town. The interior of the cup is engraved with roughly horizontal lines to mark the hours, which are numbered from 4am to midday and back to 8pm. The same line is used to indicate both the morning hour and the corresponding afternoon hour, in common with the other forms of altitude dial which we have encountered. Vertical lines on the bowl provide a form of date scale, which here is given by the signs of the zodiac.

Chalice dial by Bartholomew of Aldersbach, 1554.

Use of the dial is very straightforward. It is simply turned until the shadow of the gnomon falls on the appropriate zodiac line, at which point the top of the shadow will indicate the time of day. The markings are for equal hours, like the Regiomontanus dials and the *naviculae*. Chalice dials could also be fitted with lines marking the unequal hours, but these are rarer. The Aldersbach dial has a further scale calibrated down the chalice from zero to sixty-five, and labelled for every five. This is a solar altitude scale (for measuring the height of the sun in the sky): the chalice is turned until the shadow of the gnomon falls on this scale and the point shows the height of the sun in the sky at that moment. Such information would not be of any assistance to the user of the dial, since it was generally used for finding the time (a function already provided by the chalice) or for calculating the latitude of the observer (which would render the dial useless if it was found to be anything other than the latitude for which it was made). However, the practice of including such scales on astronomical instruments was so widespread that it is not surprising that one should appear on a sundial.

Chalice dials which could be used both when empty and when filled with water can be seen at the Germanisches Nationalmuseum in Nuremberg and at the Museum of the History of Science in Oxford. Both are dated before the law of refraction was laid down by Snel in 1621 and so must have been constructed by experiment. The Oxford example has one set of hour lines marked 'HORARIVM SINE AQVA' (hours without water) and the other marked 'HORARIVM EX PERSPECTIVA RADIOTRVM REFRACTORVM AQVAE' (hours from the perspective of the refracted [sun's] rays in water). Both these dials, however, are direction dials, dependent on the direction of the sun in the sky, rather than its altitude, and so bring us to a new era in dialling – the portable direction dial.

Chalice dial by Markus Purmann, 1599.

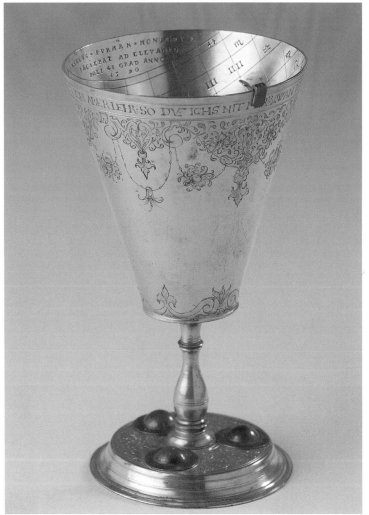

Chalice dial by Markus Purmann, 1590.

Chapter Three

THE SUN ENCOMPASSED

Introduction

The two previous chapters have dealt with altitude sundials – dials which tell the time by reference purely to the height of the sun in the sky, with the hour lines being adjusted for the season. Some of the dials we examined provide a very accurate representation of the sun's motion through the year, such as the Regiomontanus dial, the *quadrans vetus* and the *navicula*; others such as the ring dial are much more basic and sometimes do not even attempt to allow for the changing seasons. Pillar and chalice dials lie somewhere between these two extremes.

The information that we have regarding the movements of the sun means that, in addition to its height, its direction can be interpreted to indicate the time. However, we have to remember that seasonal changes affect the direction of the sun, just as they affect its height. If we were to set up a pole in the garden and mark the shadows it made at, say, 9am, 10am and 11am on 15 January, on returning to the pole six months later we would find that the new shadows we marked would be different. Not only would

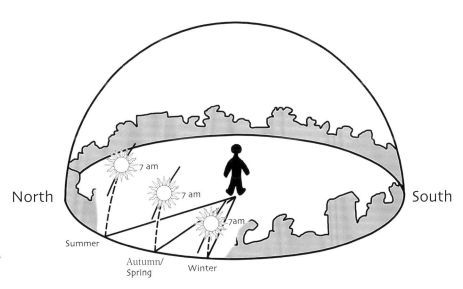

How the direction of the sun at a particular time of day is affected by the time of year.

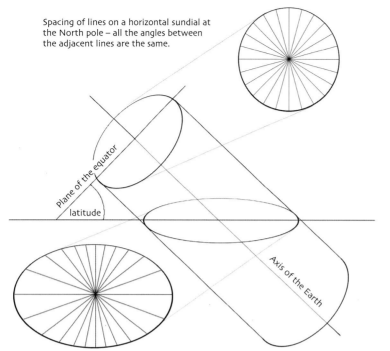

Spacing of lines on a horizontal sundial at the North pole – all the angles between the adjacent lines are the same.

Plane of the equator

latitude

Axis of the Earth

This is the pattern which appears if the 'stick of rock' is cut at an angle of 50°, and is the pattern which appears on a horizontal sundial for 50° N.

Different latitudes require very different marking of the hours on horizontal dials, as this diagram demonstrates.

they differ in height, they would also differ in direction. The only time at which the shadow is always in the same direction is midday, when the sun is always due South, and the shadow cast by it will always point North. (Of course, we have to take the effect of British Summer Time into account – in the summer the sun will be due South at one o'clock according to our watches, but this will still be midday in time measured according to the sun.)

The apparent motion of the sun is caused by the rotation of the Earth on its axis and also its orbit around the sun. If the Earth's axis were upright with respect to the plane of its orbit, a point on the planet's surface would always receive twelve hours of daylight and twelve hours of night, and the sun would appear to follow the same path through the sky on any day of the year. However, the Earth's axis is actually tilted at an angle of 23½ ° to its plane of orbit. As the Earth spins on its axis through twenty-four hours the sun appears to be rotating at an even speed around the axis and not around any other line. If you imagine the axis of the Earth casting a shadow on the plane of the equator, this shadow would also move round at an even speed on any day of the year, and so would fall in the same place at the same time, whatever the date. However, the shadow of any line which is not parallel to the Earth's axis will behave in different ways on different days of the year.

We can now see that if we want a rod to cast a shadow in the same direction at the same time every day, so that it can be used as a sundial, that rod must be parallel to the Earth's axis. This is the first step in making an accurate direction dial. However, we also know that the shadow only moves round at an even rate if it is cast on a plane parallel with the equator. For all other planes this would mean having an angled plate on which the hour lines of the dial are drawn. It is more convenient to use a horizontal plate for marking the hours, but this would mean having to work out what is happening with the hour lines. It is perhaps easiest to think about this if we take the example of a stick of rock. If the rock is marked all the way through with twenty-four lines radiating from the centre at regular intervals, and tilted until it is parallel with the Earth's axis, these lines will represent the shadows cast on a dial plate. The exposed lines at the ends of the rock will show the shadows cast on a plate parallel to the equator. However, if we require a horizontal plate, we simply have to make a horizontal cut through the rock, and the pattern of lines which we see displayed on the cut end will be the pattern of lines to be marked on a horizontal dial plate.

This is essentially the way in which a horizontal dial is prepared, although the lines are normally marked by calculations which are the mathematical equivalent of slicing through the rock. The angle at which the rod (the gnomon) is set is dependent on the observer's latitude, because the angle of the horizontal plane for a specific place is dependent on its North-South position on the Earth. The latitude is also obviously required for calculating the positions of the hour lines.

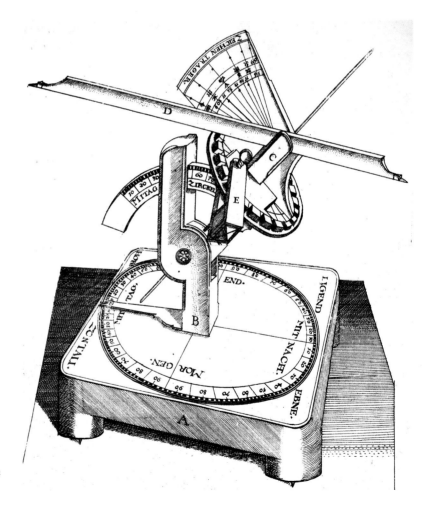

A mechanical aid for making horizontal sundials. This instrument was invented by Hans Ulrich Bachenoffen and described by him in a book written in 1627.

The theory behind this kind of dial was known from the first century AD and from then on fixed 'direction' dials began to be made and used in Europe. However, more than a millennium was to pass before portable direction dials appeared. Why was the delay so long? The most obvious answer might be that if a dial is portable it can be taken into different latitudes, thus rendering it inaccurate. However, this was not the main reason – when portable direction dials did first appear they were normally set for a particular latitude, because they were not expected to be taken far from the place for which they were made. Just as altitude dials could sometimes be adjusted for different latitudes, later chapters will show that this was true of direction dials as well.

In fact, the problem was one of knowing in which direction to point the dial, so that it could tell the right time. Returning to the direction of the gnomon, we have seen that it must be set parallel to the Earth's axis: this means that, as well as being tilted at the right angle, it must be set so that the upper end points North and the line of the gnomon runs North-South. If you are creating a permanently fixed dial there are various ways of discovering this line. You can measure the height of the sun throughout the day and mark the direction when it is at its greatest height – this direction will be due South. Or you can mark the line at night, by using the Pole Star, which indicates the North Pole. Once the line is marked, the dial can be set up on it and it will tell the correct time from then on.

However, such methods are useless if you are using a portable dial. A means of determining direction which does not rely on celestial objects is required. The reason why portable direction dials were not developed for so many centuries was because such a method had not been discovered.

Of course, the answer is clear to us now. We know that the simplest method of discovering North or South (far simpler than referring to the sun or the stars) is to use a magnetic compass. Yet the compass did not appear until the eleventh century AD. It was eventually to provide the solution to the problem of portable direction dials, but it was several centuries before it was used in that way.

The Origins of the Compass

By at least the eleventh century, the Chinese had discovered that an iron needle magnetised by the magnetic rock known as 'lodestone' would turn to point North-South if it was set floating on a bowl of water. It was used first in geomancy – divination by signs from the earth. We must remember that the hard lines which we draw between 'scientific' practices (the use of the compass) and 'religious' or even 'magical' practices (such as divination) would have been alien to the people of the Middle Ages. Divination was a means of determining future events which the Chinese believed to work and so they used it. Later, the potential of the compass on board ship for navigational purposes was realised. This latter use had almost certainly arrived by the end of the eleventh century and references to it are found in Chinese writings from the second decade of the twelfth century.

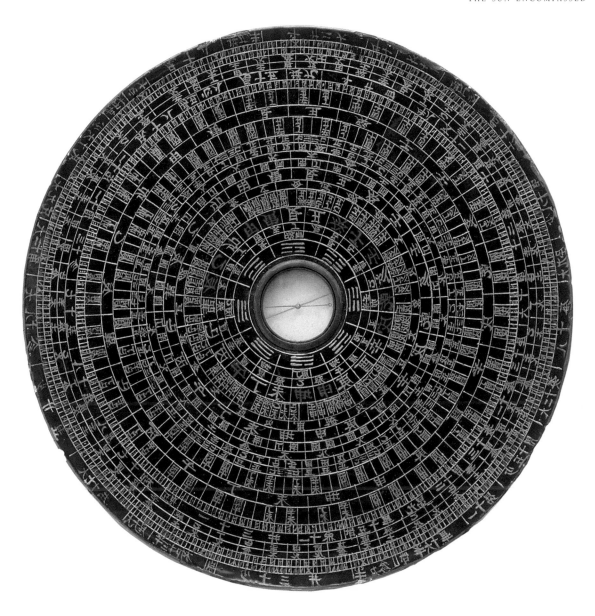

Geomancer's compass, Chinese, c.1800.

Knowledge of the magnetic compass in the West came through contact with the Islamic world, which in turn had learned of it from the Chinese. Cultural contacts between the Arabs and the Chinese began as early as the tenth century, but it was not until the following century that important contact was made between the Islamic world and the West. This new contact brought much knowledge into the West, including the properties of the magnetic compass. The earliest known reference to the compass in Western Europe is found in a treatise on natural phenomena from the closing years of the twelfth century.

Use of the compass became widespread during the thirteenth century, and it was Europeans who first devised the 'dry' compass by suspending the needle on a pivot rather than floating it in water. However, it was probably not until the late fourteenth century that it was discovered that a compass added to a horizontal direction dial would make that dial portable.

Applying Compasses to Sundials

The horizontal direction dial had several advantages over the altitude dial. Significantly, unlike altitude dials, these new dials were provided with separate hour lines for each hour of the day so they did not suffer from the same problem as the altitude dials, for which the user had to know whether it was before or after midday. As long as the compass was correctly oriented and the dial plate was held horizontally, the time would be shown directly by the shadow of the gnomon among the hour lines.

The other main advantage of direction dials was that they were easy to make. Marking out the lines could be done quite simply once the angles had been calculated for each hour. Straight lines were easier to draw than the complicated curves needed for pillar and ring dials, while the mathematical expertise required for the preparation of Regiomontanus dials and *naviculae* was far greater than was necessary for making a simple horizontal dial.

The ease of use and production of dials with compasses meant that they rapidly gained popularity, becoming by far the most common form of sundial for rich and poor alike. Horizontal dials made for the the rich continued to be manufactured from costly materials such as brass, silver and ivory. However, replacement of these materials with wood produced a cheap variety which could be afforded by the lower classes of society.

Various types of horizontal dial survive from the period in which they were first produced. By the middle of the sixteenth century horizontal dials had become as uniform in design as pillar dials and ring dials: however, before then the variety of different types was large. Some appeared as cruciform dials (like the one shown here, from the British Museum); some were made of wood, some of metal, some even of stone. Two forms which did gain a certain amount of popularity probably mark the two ends of the market for these instruments: the signet-ring dial and the simple wooden dial, and it is these types which will be considered in the first two examples of this chapter.

Cruciform dial, possibly Italian, possibly seventeenth century. The compass is missing from this dial, but would have fitted into the circular depression at the centre of the body.

Signet-ring Dials

The sundial shown here is a fairly late example of the signet-ring dials produced for the rich. It is also a lavish example, made in gold and black enamel with very ornate acanthus decoration. The signet itself carries the letters 'MIMR' in the centre and two further letters (probably H and R) at the sides, but I have been unable to work out the meaning of these initials (although they could be the initials of the owner). The signet opens up to reveal a tiny sundial beneath, with further engraving (two more initials – H and C – and stars and a central flower) inside the lid. Hour lines are marked on the dial face from 5am to 7pm with a tiny compass placed at one edge. The gnomon is missing but would almost certainly have been provided by

Signet ring dial, possibly Swiss, c.1570.

Finger ring dial, probably German, sixteenth or seventeenth century.

a thin string which stretched taut from the meeting point of the hour lines to the top of the lid, when the dial was open, and folded up underneath the lid when the dial was closed. The hour lines have been drawn for a latitude of 48° N, which corresponds with areas central to Northern France, Switzerland and Southern Germany. It has been suggested that the dial was made in Switzerland. There is no date, but we can guess that it was made in about 1570, since a very similar object (also marked 'HC') in the Whipple Museum in Cambridge carries the date 1568.

The contrast between this dial and the type of dial mentioned in the previous chapter is very striking. The small size of the ring deprives the dial of any kind of accuracy, whereas the Regiomontanus dial, the *navicula* and the *quadrans vetus* are examples of highly accurate dials where the progress of the sun through the sky has been carefully mapped onto the lines laid out on the dials. Those dials would have been able to tell the time to the nearest quarter of an hour, but this signet-ring dial would be unlikely to have an accuracy greater than the nearest hour. These dials were probably designed as expensive gewgaws or playthings rather than as useful time-tellers. The dial is simply a gimmick, hidden beneath the signet ring, and

perhaps only displayed as a marvel. One could perhaps compare them to a credit-card-sized computer with a keyboard too small to use and no function other than the display of how much technology can be fitted into a tiny space.

Signet ring dial, possibly Swiss, c.1568.

Dials for the Poor

In comparison to the relatively large number of expensive sundials which have survived from the Middle Ages and the Renaissance, there is a dearth of the ones made for the lower classes. There are various reasons for this unequal preservation. Expensive objects tend to be carefully conserved and handed down within families, becoming heirlooms and providing links with the past. Moreover, the dials designed for the rich were made from durable materials which survive the passage of time well. In contrast the cheap dials were made from cheap materials and were thrown away when they had reached the end of their useful life – if, for instance, the hour lines had rubbed off the dial plate. Many of these cheap dials were made of wood, which decomposes rapidly unless preserved in particularly favourable conditions.

However, wooden dials have survived in some cases and so we do have some idea of the sorts of dial which would have been available for the less well off. A group of very simple wooden sundials was recovered when Henry VIII's ship, the *Mary Rose*, was raised from the seabed in the 1980s. The mud of the Solent had saved the dials, along with the other contents of the ship, from decomposition, and they were in such good condition after more than four hundred years under the sea, that it was possible to create accurate replicas of the originals.

The Treasures of the Mary Rose

The *Mary Rose* was originally built at Portsmouth in 1510, on the orders of Henry VIII, being named after the King's favourite sister (the 'Rose' probably refers to the Tudor rose, the emblem of the royal family). In her original form the *Mary Rose* may have followed old designs of shipbuilding with a hull constructed from overlapping planks within which the beams and frames were set, and perhaps only a single mast. However, in 1536 she returned to the dockyard and was reconstructed, following new plans, as a three-masted warship of the type which had become common throughout Europe in the fifteenth century, but which had only reached Britain during Henry VIII's reign. The refit also allowed the number of her guns to be

The Mary Rose as depicted on the Anthony Roll of 1546.

increased from seventy-eight to ninety-one and the arrangement of the guns was altered, with artillery set on the main deck and firing through gun ports in her sides. She is the earliest known example of an English warship with such an arrangement of guns, which continued to be used until the 1850s, the most notable example probably being that of HMS *Victory*.

On 19 July 1545, the *Mary Rose* left Portsmouth as part of a fleet sent out to engage a French force in the Solent. As the King looked on, she heeled over abruptly while going about and sank within a few minutes. The rapidity of the sinking may very well have been assisted by the guns of the main deck running loose from their constraints and also by water pouring in through the new gunports on the same deck. Other contributing factors may have been the enemy action and also the fact that she was carrying about three hundred soldiers, thus nearly doubling her usual complement of men. There were only thirty-five survivors.

An immediate attempt was made to raise the *Mary Rose*. However, because she had sunk so rapidly she had embedded herself firmly in the sediments of the seabed, and the attempt only succeeded in tearing the mainmast away from its housing. Thereafter, salvage was confined to the contents of the ship, with the yards, sails and some guns being recovered. The *Mary Rose* was then abandoned in the mud for nearly three hundred years. It was not until 1836 that she was rediscovered by John and Charles Deane, professional divers who managed to recover fifteen guns and some other material. After this brief activity, there were no further attempts until the 1960s when the wreck was discovered once more and plans were made for the complete recovery of the ship. During the 1970s the excavation of the packed contents of the ship proceeded apace, and many important finds were made. These included common items such as coins, spoons, stone, iron and lead shot, the guns themselves (some still mounted on the earliest

Pocket horizontal dial recovered from the Mary Rose, sixteenth century.

*Twentieth-century replica horizontal dial
copied from an item found on the* Mary Rose.

gun carriages yet known), candlesticks and combs, and the chest of the bar-
ber-surgeon, still complete with its contents. A chest in one of the officers'
cabins contained some navigational instruments including dividers and the
oldest ship's compass known, as well as one of the small wooden sundials.
The other sundials were found in various places throughout the ship. The
hull itself was finally raised in October 1982, and the *Mary Rose*, along with
her contents, can now be seen at the Portsmouth Dockyard.

The presence of these sundials on board the *Mary Rose* might seem rather odd – would they have been useful as ocean-going instruments? The pictures show one of the original dials, and a replica which was made from it in the 1980s. The dial is very simple in form, marked only with lines for the hours and making use of a folding brass gnomon. A small compass is set into the dial plate with no markings other than a North-South line. The decoration is simple and only in two colours – red and the black which was used for all the other markings on the dial. The smallness of the dial limits its accuracy, and there were much more accurate timekeepers on board the ship: the sandglass would have been the main time indicator, turned every half-hour, its turning marked by the sounding of the ship's bell. These tiny dials could not compete with the accuracy of the sandglass, but perhaps individual sailors might have found them useful on land rather than at sea. Even so, they are only accurate to the nearest hour.

Another problem with the usefulness of these dials relates to the fact that they were made for specific latitudes. As I mentioned earlier in this chapter, dials like these cannot be adjusted for different latitudes and will rapidly become inaccurate if they are taken too far North or South of their place of manufacture. Yet sea voyages will inevitably take a ship through different latitudes. Moreover, while some of the dials which have been retrieved from the *Mary Rose* were made for the approximate latitude of Portsmouth (roughly 50° N), there were others which were set for 49° N, and were probably made in Nuremberg or Augsburg (later to become important centres of dial manufacture). These dials were already far enough North of their intended place of use to begin to add further inaccuracies to those created by the small size of the dial plate and the compass. It is possible that a sailor using one of these might not be able to read the time more accurately than to within two hours.

So we are left with the unanswered question of what use these dials might have had on board the *Mary Rose*. It is one which is almost impossible to answer, although it does raise further questions concerning the need to know time accurately. Perhaps for a sailor on land at any rate, a margin of two hours was good enough, and he carried his dial with him, knowing that it would provide him with a sufficiently accurate measure of time anywhere within the waters of France, Holland and Britain, to which he was most likely to be called in the line of duty. Or perhaps they were kept by seamen who were ignorant of the way they worked, and merely saw the dials as status symbols, just as signet-ring dials were status symbols for the rich.

A Cautionary Tale

For my last example of horizontal dials, I want to look at a slightly more unusual dial for the period. Unlike the simple wooden ones produced for sailors and people of similar standing, dials made for the rich were often elaborately ornamented and this one, at first glance, appears to be no exception. It is perhaps unusual in being made from marble – stone tended to be reserved for fixed dials. The brass gnomon carries a depiction of the moon

Underside of 'Sabeo' dial showing the anachronistic epact table.

Horizontal sundial, early twentieth-century, inscribed as made by Berardino Sabeo in 1552.

as a grotesque and a similar style is used for the engraving of the sun around the base of the gnomon. Lines are shown not only for the hours, but also for the half, quarter and eighth hours. Around the hour scale the eight winds are marked in Latin and a motto is engraved at the centre of the dial: *Horas non numero nisi serenas* (I only count the bright hours). The under side of the dial carries an instrument for converting moon time to sun time, as well as a table for finding the date of Easter.

While the dial itself is relatively unusual, additional interest is aroused by an accompanying manuscript giving details of the original owner and some of the other former owners, unusual for an object of such age. The dial itself carries the basic information about the presumed original owner, because a dedication has been engraved on the sides of the base. This reads *Bernardinus Sabeius. f. Patauii 1552 ad Lat. 45.° & dicauit Amico suo caris. Jo Paduanio Ver. Prof. Clar.*, which is translated as 'Made by Bernardino Sabeo of Padua 1552, for latitude 45°, and dedicated to his most beloved friend, Giovanni Padovan, professor at Verona'. The manuscript gives further information about Giovanni Padovan, who was Professor of Mathematics in Verona, and a 'most learned Astrologer'. The manuscript also provides details of the descent of the sundial through the family until it came into the hands of Francesco Padovan. While he was a student at the University of Padua he gave the instrument to his teacher, Giuseppe Zouldo, who was the Professor of Astronomy. Zouldo was the author of the manuscript, which is dated 10 June 1789.

When I first examined this dial and its manuscript I was very interested to discover such solid details about the maker and the early owners of the instrument. However, when I looked at it more closely, several things seemed a bit strange. The hour of four o'clock is normally written as 'IIII' on dials, but 'IV' appears on this one, though this might have been simply an unusual quirk of the maker. The appearance of a motto is very unusual on an early dial (they were a favourite addition of nineteenth-century dial makers), but this could once again be put down to an idiosyncrasy of the maker. The feature which really attracted my attention was the table of epacts on the reverse (a table used for determining the date of Easter), which is described as being listed 'according to the Gregorian Calendar'. The Gregorian Calendar was introduced by Pope Gregory XIII in 1582, in place of the Julian Calendar (named after Julius Caesar) which had got out of step with the orbit of the Earth. Since the Gregorian Calendar was not introduced until thirty years after the date inscribed on the dial, the date of 1552 cannot be genuine, and this, linked to the other problems and the general style of the markings, points to the instrument being a fake, probably manufactured at the end of the nineteenth century or the beginning of the twentieth. So, while the inscription on the dial and the manuscripts in the case all provide supporting evidence for the authenticity of the dial, careful study of the dial itself betrays the telltale marks of the forger. Forgeries of sundials are fairly common, and the collector who does not want to be duped must be careful to learn the signs (like the ones in this case) which indicate a fake.

As I have said, the appearance of the maker's name on a dial of this period would have been unusual. It was only with the establishment of workshops dedicated to the production of sundials and related mathematical instruments that makers' signatures began to appear on dials with any regularity. The next chapter will turn to the introduction of these workshops and the beginning of a more regular trade in sundials.

Chapter Four

MULTIPLICATIONS

Introduction

Makers' names did not appear regularly on dials until the sixteenth century, and only then does the opportunity arise to study the makers themselves and not just the objects they produced. This new area of interest appears just at the point when few major design changes were made, in mathematical terms, and the focus therefore moves away from new ways of using the sun to show the time. This chapter will pick up the narrative of social history – studying makers and their place in the society of the time, and look at the wider areas which affected the development of the workshops rather than the users of the instruments.

Workshops for the production of sundials (and of scientific instruments in general) only began to appear in large numbers during the sixteenth century. Of course they had existed before this but it was only really at this time that the trade of instrument maker became well established. One of the reasons for this may have been the increased market for navigational instruments, which the advent of the compass and the beginning of major voyages of exploration had created. The establishment of a mass market for compasses was particularly important; often the same makers manufactured dials containing compasses. These types of dial became popular during the sixteenth century as more and more makers produced them, and Nuremberg especially became a centre for the trade in sundials.

We will look first at those Nuremberg sundials and their makers. However, there were two other places which were of some importance during this period, and they will also be discussed. Those places are Augsburg and London, and they are important because of individual makers (Christoph Schissler and his son in Augsburg, and Humphrey Cole in London), unlike Nuremberg where there was a large number of different makers and extensive family businesses. The types of dial in all cases are different from the fairly simple direction dials which we looked at in chapter three. Almost all the sundials produced by these makers were multiple dials, with a selection of other dials being joined together into one compendious instrument. We will see how this was realised in different ways as we look at the different areas in which the dials were produced.

Nuremberg and the Trade in Ivory Sundials

During the second half of the thirteenth century Nuremberg rose from the relative obscurity in which it had been wrapped before that point. The immediate reason for its growth was its establishment as a free imperial city and a member of the League of Rhineland cities. The privilege of being an imperial city ensured that the city was not dominated by a local prince but was able to grow as a major centre and to extend its possessions and interests. This was particularly true in the area of trade, and Nuremberg benefited from the movement of trade away from the Hanseatic ports of Hamburg and Lübeck, among others. Various wealthy merchant families in the town were instrumental in creating a trade network which stretched far across Europe. This network was further bolstered by Nuremberg merchants in Italy and Flanders, and by the city's favourable attitude to foreign merchants wishing to settle and conduct their trade in the city. Nuremberg was also favoured by the expansion of overland trading routes in the late fifteenth and early sixteenth centuries. Lastly, it was well placed for travel between the Mediterranean and the Baltic Seas and between Spain and its colonies (the Hapsburg lands of Austria and Germany). The main commodities imported by the city were cloth, fruit, and spices, while there was a steady outward stream of high-quality manufactured goods made of glass, metal, leather, paper and clay.

The growth of Nuremberg as a trading centre led to the establishment of numerous workshops for the manufacture of highly crafted objects. Many people had fled to the city during the Black Death in the fourteenth century and the resulting abundance of skilled workers brought craft specialisation and a vastly increased productivity. In particular the crafts based on metalworking played an important part in the economy of the city and Nuremberg was famed throughout Europe for the quality of its metal goods. Gold, silver and iron items were all produced in large quantities, but it was brass which was the favoured material. The copper, tin and zinc ores required for its production were purchased through wholesalers from mines in the Harz, the Upper Palatinate and the Black Forest. Early brass goods included weapons, bells and domestic goods.

There is evidence that the manufacture of brass scientific instruments in Nuremberg began as early as the middle of the fifteenth century. Many of these instruments were probably designed and commissioned by mathematicians and natural philosophers, but the more common types (such as dividers and rules) would probably have been produced in large quantities and kept in stock by the makers. Other common instruments whose manufacture was wholly overseen by the craftsmen were magnetic compasses and sundials (manufactured in the same workshops, since sundials required compasses and so the makers learnt the techniques for producing both types of instrument). Compass makers were known in Nuremberg from at least the 1480s. It may seem odd that a landlocked city should have been one of the major producers of compasses, which were largely marine instruments. However, the extent of the trade net centred on Nuremberg ensured that the compasses produced in the city could be readily transported to

View of Nuremberg, *engraving by*
Wenceslaus Hollar, 1635.

ports all over Europe. The fact that the city was such an important centre for trade is reflected in the sundials, which were mostly designed for use in several latitudes (like the Regiomontanus dials and *naviculae* of earlier centuries), and so would have been of use to the merchants and nobility who travelled regularly throughout Europe.

The Compassmakers of Nuremberg

Between 1550 and 1700 the craft of dial-making was kept within a very limited number of Nuremberg families. The names which crop up on dials from the city are found time and time again: Karner, Lesel, Miller, Reinmann, Troschel, and Tucher. There were various connections between the families: Elisabeth Karner, the widow of Conrad Karner (fl.1564–c.1585) married Albrecht Lesel in 1590 and they probably also had blood links with the Reinmann family, since Michael Lesel used the same maker's mark as the Reinmanns. The Karners were probably the most prolific family, producing dials from the sixteenth century right through to the eighteenth century, but all the families produced a large number of dials. Instrument making in early modern Nuremberg really was a family affair, and contrasts very strongly with today's big firms where continuity within one family is very uncommon.

Ditpych dial by Thomas Tucher, mid-seventeenth century. Exterior, showing wind rose.

Diptych dial by Hieronymus Reinmann, 1556. This view shows the equinoctial dial and the table of latitudes on the outside of the dial.

The speciality of the Nuremberg dial makers (or compassmakers, as they were known, because of the development of this type of dial from compasses) was the ivory diptych dial. The use of the term 'diptych dial' referred to the form of the dial, which consists of two leaves or tablets hinged together, forming a right angle when open. They are similar to diptych paintings where two paintings are hinged together, often with further paintings decorating the outer sides of the leaves. The use of ivory in such great quantities is unusual in dials, being found otherwise only in the magnetic azimuth dials produced in Dieppe during the seventeenth century (see chapter six). Most of the instruments were decorated with brass, which was used for hinges and clasps and various other parts of the dial, and the faces of the dials were normally brightly coloured in red, blue, green or even gold.

The main dial on a diptych was usually a horizontal direction dial on the inside of the lower leaf, with the compass set at one end. These were fairly similar in essence to the earliest form of portable direction dial. However, the presence of the upper leaf allowed space for the inclusion of other dials, not just a cover for the main horizontal dial. A string set between the inside of the two leaves would serve not only as a gnomon for a horizontal dial on the lower leaf, but also as one for a vertical dial on the inside of the upper leaf. Strict definitions of the term 'diptych dial' have often insisted on the

Diptych dial by Leonhard Andreas Karner, 1733. The inside of this dial has a vertical dial and a horizontal dial which use the same string gnomon. Below the horizontal dial is a dial marked out for Italian and Babylonian hours.

presence of these two dials – vertical and horizontal – on the inner surfaces, and have ignored dials which are diptychs in the sense of containing two leaves, but which do not have the vertical dial on the upper leaf.

The other standard direction dial to appear on the diptychs is an equinoctial dial, on the outside of the upper leaf. The name of this dial comes from the way in which this dial is set up to tell the time: the plate which is marked with the hours is set so that it is parallel to the equator – so the dial is referred to as either 'equatorial' or 'equinoctial'. The correct angle for the plate is set by a strut positioned between the upper and lower leaves, with one end sitting on a latitude scale. The gnomon of the equinoctial dial is supplied by a small pin, normally stored in a hole in the side of the lower leaf, and placed in the centre of the equinoctial dial, at right angles to the leaf.

These were the main parts of most diptych dials, but they were not the only instruments which were included on many examples of the type. Diptych dials were in some ways like some modern digital watches – they could perform various other operations than simply showing the standard time in a particular place. The pictures show some of those other functions. In the image at the top of page 50 we see a wind rose, marked with the directions and the names of the eight major winds: a wind vane could be set in the hole at the centre of the ring and the dial placed outdoors so that the direction of the wind was indicated by the vane.

The picture here shows the interior of a dial by Leonhart Miller which includes three further common features of the diptych dial. At the top of the upper leaf we see a dial which can be used, with a horizontal pin gnomon, for measuring the hours of daylight and also for indicating the sun's place in the zodiac. Below it is a table of major European cities with their latitudes: this table provided the user with the latitude for the place in which he was using the dial, and so allowed him to set the dial correctly. As I mentioned in the previous chapter, horizontal direction dials are made for specific latitudes and so can only be used in a restricted number of places. However, if hour lines for different latitudes are marked on the dial plate (as here) and the string gnomon can be adjusted (by moving the top of the string to a hole in the upper leaf which corresponds with the latitude) the dial can be transported to different latitudes and still tell the time accurately – the range of latitudes in which this particular dial could be used was from 42° N (the latitude of Rome) to 54° N (Northern Germany). The final feature which appears on the dial is a *scaphe* (bowl) dial set into the lower leaf. This dial does not mark ordinary time but two other time systems that were used during the Middle

Diptych dial by Leonhart Miller, 1628. Included on the inside of this instrument are a dial for showing the length of the day, a table of latitudes, a horizontal dial for several latitudes and a dial for marking Italian and Babylonian hours.

Diptych dial by Michael Lesel, c.1612.
Exterior, showing epact tables.

Ages and the Renaissance – Italian hours and Babylonian hours. Italian hours (which were still used in Italy until the eighteenth century) were reckoned from sunset and numbered from zero to twenty-four; Babylonian hours were similar but were reckoned from sunrise rather than sunset. These two hour systems were divided into equal hours, not the unequal hours which we met in earlier chapters.

The photograph above shows the last common feature of the diptych dial: the lunar volvelle. This device allowed the dial to be used on nights when the moon was bright by providing a table for converting the time shown by the moon to the actual time of the night. The volvelle was often surrounded by two epact tables, used to determine the date of Easter. The reason for having two tables was because of the two different calendars used in Europe from the end of the sixteenth century. The Gregorian Calendar was introduced by Pope Gregory XIII in 1582 and was immediately adopted in most of the Catholic countries of Europe. However, Protestant countries were slower to move away from the Julian Calendar and so epact tables for both calendars are common on these dials until well into the seventeenth century.

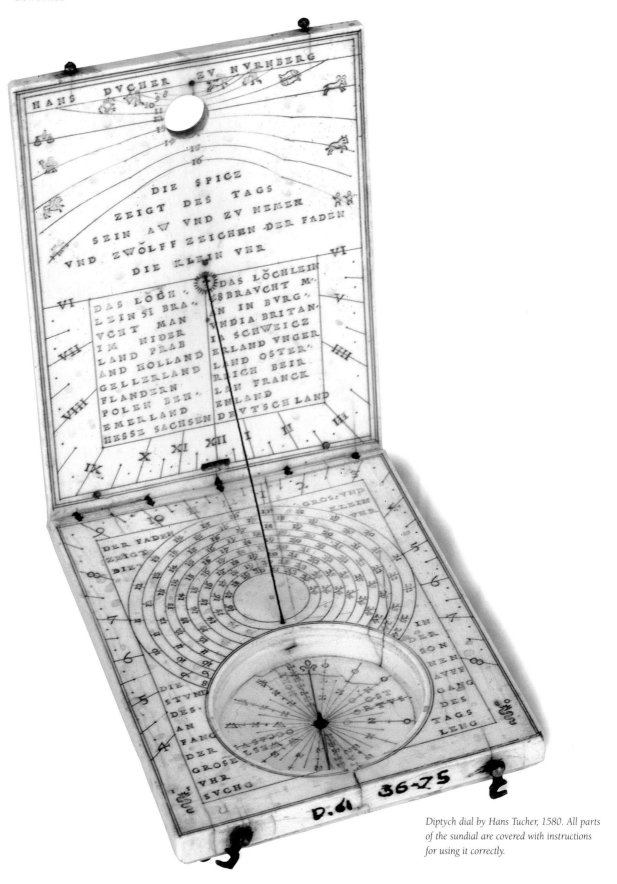

Diptych dial by Hans Tucher, 1580. All parts of the sundial are covered with instructions for using it correctly.

A Simple Diptych Dial by Hans Tucher

The detailed example which I have chosen is a dial made by Hans Tucher (or Ducher) in 1580. The dial is signed at the top of the upper leaf on the inside and is also marked with Tucher's maker's mark – a crowned snake – at the bottom of the lower leaf on the inside. A further mark appears on this leaf at the bottom: a Gothic script 'N' for Nuremberg. This was the mark which a master craftsman set as a sign of approval on any dial which was deemed to be of sufficiently high quality.

This dial is rather different from the examples which have been shown earlier. While it includes many of the standard features – vertical and horizontal dials, a dial for marking daylight hours and a lunar volvelle – it is remarkable for the simplicity of the decoration and the limited use of colour (all the engraving on the dial is in red). The outer surface of the upper leaf carries a series of horizontal dials rather than the more normal equinoctial dial. This dial carries scales for various different latitudes from 42° up to 54° N. The horizontal dial which appears in the standard place on the lower

af is unusual in that it only has one standard hour scale, marked out ound the edge for 45° N. The other hour scales are a set of scales for alian hours, which would be used at different seasons of the year for reading the time as hours reckoned from sunset. The outermost scale would be sed for the time around midsummer when there are most daylight hours; he innermost scale would be used during late December and early January, ind the other scales would fit into the months between the two solstices.

These unusual features apart, the most interesting part of the dial is the inclusion of an 'instruction booklet' on the various surfaces of the dial. The inscriptions which can be seen in the pictures are explanations of the various parts of the dial and also lists of the places for which the different scales of the horizontal dial on the upper leaf are required. The inscription at the top of that dial gives the title and immediately runs into the first part of the 'latitude table': 'An instruction concerning this dial in which the latitudes of various places are noted. The hole 54 is required for England, Scotland, Lifland [an area of Southern Scandinavia], Moscow, Iterland [I have been unable to identify this place], Zealand [the island part of Denmark], Sweden, Norway, Denmark.' Lower down the user is told that to use the dial he must 'thread the string through the hole of the latitude which you require'. Inside on the upper leaf the explanation of the use of the dial for daylight hours and the vertical dial is given: 'the point [i.e. of the horizontal pin gnomon] shows the increase and decrease of the day and the twelve signs [the zodiac]; the string shows the common hours'.

These instructions and lists of latitudes continue on all the faces of the dial. So Tucher has ensured that the owner of this dial will have a reminder of its use always to hand, as well as the necessary table of latitudes if using the dial while travelling. Other dials by Tucher include similar instruction booklets, but this one is considerably more detailed than most of the others known.

View of Augsburg, *engraving by Wenceslaus Hollar, 1635.*

Augsburg

While the trade in ivory sundials was thriving in Nuremberg a rather different style of sundial was being produced in Augsburg, another important Bavarian town. Like Nuremberg it was an imperial town, but unlike Nuremberg it had few old patrician families, the upper echelons of the city class structure being rather more dominated by groups of financier-merchants, of which the most important were the Fugger family. The Fuggers derived their wealth from colonial imports (particularly spices, silks and woollens from Venice) but their central role was as financiers to the Emperor and his family, and as papal bankers for much of Germany, Scandinavia, Poland and Hungary. The association with the Imperial Court had both benefits and negative effects. On the positive side, the Fuggers were granted large gifts of land, such as the Hapsburg silver and copper mines in the Tyrol, in return for ever-increasing loans. However, since no one could sue the Emperor for non-payment of loans, the Augsburg bankers found themselves (and along with them the other inhabitants of the town) dependent on keeping the goodwill of the Hapsburgs.

This state of affairs led to complications when the teachings of the Reformation reached Augsburg. During the 1520s the guilds and other representatives of the populace sided in favour of Lutheran doctrines, placing themselves in conflict with the merchant bankers, who required the favour of the Catholic Emperor to retain their wealth and their standing. Until 1530, the Fuggers and other leading families attempted to keep the city from forming close links with newly Protestant princes and cities, with some success. However, the strength of public opinion eventually led to Augsburg joining the league of Protestant states. When civil war swept the German states in 1546–47, the Protestants were defeated by the Emperor's forces, and Augsburg was forced to submit to harsh changes. In particular, the guilds were stripped of most of their power, being subjected to a council of Catholic patricians, who followed Charles V's orders, and restored the rebellious town to his rule.

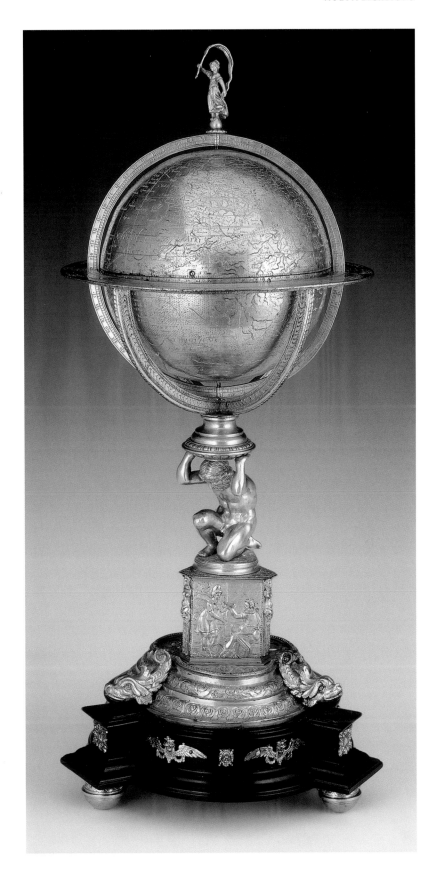

Terrestrial globe by Christoph Schissler, 1597.

The Augsburg Makers

Unlike Nuremberg, with its group of important dial-making families, the fame of Augsburg as a source of scientific instruments in the sixteenth century rested mainly on the work of two men – Christoph Schissler and his son Hans Christoph, whose work was known for its combination of technical accuracy with elegance of design and decoration. The elder Schissler was born in Augsburg in about 1531. He was clearly apprenticed to an instrument maker and became a master in 1553, the year in which he also married. He remained in Augsburg throughout his life and was known as a prolific maker of astronomical instruments, sundials and surveying instruments. He did, however, make several major journeys to deliver commissioned instruments or to make new and better ones. These journeys brought him to the courts of Kaiser Rudolph II (in Vienna and Prague) and of August, the Elector of Saxony (in Dresden). Rudolph was so impressed with Schissler's work that he presented the maker with a silver-gilt drinking vessel, on the occasion of his golden wedding. Schissler's surviving works include sundials, compendia, astrolabes, quadrants, armillary spheres, globes, dividers, compasses and other drawing instruments. His finest period seems to have been between 1568 and 1579. After the death of his wife, Euphrosina, in 1603 he remarried the following year. He died in September 1608, his second wife, Katharina, outliving him for at least twelve years.

Schissler left two sons and two daughters. The eldest son, Hans Christoph, was born probably in the late 1550s. He worked in his father's shop in Augsburg, although he leaned more towards clockmaking than instrument making. He married Katharina Scheblin in 1580, continuing to assist in his father's workshop until about 1585, when he began to spend most of his time in Vienna. From 1591 he extended his work to the Emperor's capital at Prague; he served at the court there with some brief commissions at the imperial residences in Dresden and Vienna. He was still working in 1610 when he was established as a turret-clock maker, and was still alive in 1626. Very few of his instruments survive (a few sundials and some drawing instruments) and they all demonstrate the strong influence of his father.

A Schissler Compendium

As I have said, Christoph Schissler was one of the best instrument makers of the Renaissance, and this astronomical compendium is a fine example of his work. The astronomical compendium was a new form of dial which was much favoured among the wealthy between about 1550 and 1650. Like the diptych dial it contained more instruments than a single dial, and was designed to fulfil a whole series of uses rather than the single one of telling the time by the sun. Most compendia were made of costly metals – commonly gilt-brass, but occasionally silver – and were clearly only intended

Astronomical compendium by Christoph Schissler, 1575, showing vertical dial.

for members of the aristocracy, unlike the diptych dials which would have been affordable by the merchant classes as well.

This particular compendium is signed by Schissler and dated 1575, during the period in which he was producing his finest works. The lid carries a typical table of latitudes, with Southern cities listed on the outside and the towns of Northern Europe appearing inside. The extent of the table is unusually wide, stretching from Sardinia and Granada in the South to Lübeck and Stettin in the North, and from Lisbon in the West to Vienna in the East. All names are given in Latin. The reverse of the compendium carries a very fine vertical dial engraved with the two-headed imperial eagle.

The main horizontal dial is supplied with a series of three plates which can be placed above the compass and used with an adjustable string gnomon. Each plate has a set of hour lines on each side, so allowing six main

Interior of the compendium showing the compass and the three horizontal dial plates.

latitudes to be covered (from 39° N to 54° N). Each plate is decorated with three coats of arms: it was thus clearly made for an aristocratic patron, perhaps a pope, since one of the coats of arms carries the papal crown and the cross keys of St Peter. We have seen that Schissler had connections with the imperial court in Prague, and it may have been through these links that this compendium was commissioned.

The plate on which the compass is set is decorated with different colours of enamel, and also with four fine vignettes of rural life. The choice of subjects – sheep-shearing, harvesting, wine-making and a man warming his feet at a fire – suggests that the engravings were intended to symbolise the four seasons of the year. The reverse of the compass plate is again finely enamelled in red, green and black with various ornamental floral patterns and a bird on the compass itself.

Perhaps the most interesting part of the compendium is the map which is carried on the inside of the back of the instrument, within the markings for a horizontal dial. The land is shown in gilt-brass on a sea of silver, with ships also shown in brass. It is not immediately recognisable but the numbers marked at various points on the map correspond with a table above it listing places in Europe and the Mediterranean. One might think at first that while Schissler was a master instrument maker his knowledge of geography was poor and resulted in this rather warped representation of Europe. However, a number of his other compendia carry detailed and accurate maps of Europe (see the picture on the next page), so it is clear that there must be another reason for the particular form of map which appears here.

What is actually happening is that the map is used for another purpose than the standard cartographical representation of Europe. This is indicated by the presence of the scale of a horizontal dial around the outside of the

The 1575 compendium showing the non-standard map of Europe.

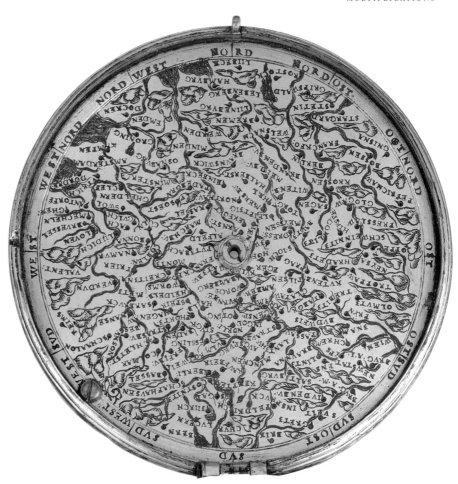

Astronomical compendium by Christoph Schissler, 1560. Inside of lid, showing map of Europe.

When the map from the 1575 compendium is stretched to the normal representation, this is the outline which appears (shown in black). An outline of Europe is shown beneath it in red for comparison.

map: the two were intended to be used in conjunction. The map is laid out so that if the dial is used in the longitude for which it was designed (that of Augsburg) the shadow of the gnomon will not only show the time of day for that place but also the places at which it is midday. So, for instance, at 1pm in Augsburg the dial will show that it is midday in Paris and Toulouse (both of which lie on the one o'clock hour line, in the way in which they are marked on this map). There are places where the map has not been constructed quite accurately – Corsica and Sardinia are shown as lying east of Augsburg when they are in fact west of it – but on the whole it would have performed its function well.

London, *engraving by Wenceslaus Hollar, 1666. These two views show the same scene before and after the Fire of London.*

London

While the instrument-making trade was quite firmly established in mainland Europe by the middle of the sixteenth century, developments were rather different in England. It is well known that the Renaissance came to London at a much later date than to other European centres, and other cultural growth was correspondingly slow. London in the mid-sixteenth century was only just beginning to become a major city: wars within the country (the Wars of the Roses) and with Europe had put a heavy strain on the nation's resources. However, in the latter years of Henry VIII's reign and during the long rule of Elizabeth the country stabilised and England began to grow both economically and culturally.

Throughout the sixteenth century London dominated the development of trade and commerce in the English nation. It was by far the largest city in the kingdom, rising from 75,000 people in 1500 to 220,000 by the end of the century (the second largest city was Norwich, and this could only boast 29,000 inhabitants by 1650). This enormous increase in numbers was due not only to an expanding population but also to the attraction of the capital to those in other areas of England. Much of the country's wealth was concentrated in London, and it was here that the wealthy merchants made their fortunes, with ninety per cent of overseas trade passing through their hands. Those who were able to reach master's status in the guilds were generally guaranteed security for their future and the prospect of further advance for their families.

The government of London was largely autonomous, being controlled by the great merchants and bankers of the city, rather than the crown. The relationship between monarch and capital was always a stormy one, although the Tudor period was largely an amicable time (apart from Mary's reign, when the citizens stolidly affirmed their Protestant leanings). The city certainly gained importance under Henry VIII, when new dockyards were established in Deptford and Woolwich. However, real prosperity did not arrive until the less troubled years of Elizabeth, in the second half of the century. At that time trade flourished, theatres opened in many areas of London, musicians found themselves in great demand at the court, and luxuries such as carpets, pillows, good food and elaborate clothes began to be available to a significant proportion of the populace.

The Guilds and the First Instrument Makers in Britain

By the sixteenth century the guild system was well established in London. There were twelve great companies, who controlled much of the city's government and trade, and a host of smaller guilds. Each guild was responsible for training apprentices in the craft and for ensuring that goods produced in the workshops were acceptable for sale as guild products. An apprenticeship normally lasted seven years, after which time the young craftsman had the right to practise as a journeyman, and eventually to become a master on payment of a fee to the guild. At least, that had been the original process: by the time of the Tudors it was increasingly hard for newcomers to gain freedom within the companies and to become master craftsmen. Not only could freedom of the company be obtained through apprenticeship, it could now be passed down from father to son, or could even be bought, as long as the applicant was over the age of twenty-one.

Apprentices came mostly from the large number of artisan families already involved in trade in London. However, it was also common to find fatherless boys being pressed into the trades, or for yeomen from the surrounding counties to send their sons to London to learn a craft which might lead to an enhanced standing in the long term. Even the younger sons of gentlemen were sometimes sent to London to learn a trade for their livelihood: the rank of merchant, despite the association with manual work, was not one which was necessarily looked down upon by the gentry of this period, since the more important merchants in the city had considerable power and influence in the capital.

At the end of his term, an apprentice could gain his freedom if he had the consent of his master and the approval of the court of the guild, and was prepared to pay for entry into the guild. Many apprentices opted to remain in their masters' service for a further period, moving up to the rank of journeyman, whether they had taken their freedom or not. While the position of freeman in a guild had its own privileges and opportunities, it also carried burdens and obligations, and many artisans were loath to take those responsibilities upon themselves until they were sure of being able to establish themselves sufficiently well to earn a living. Many who had no desire to own their own workshop were happy to remain in the relatively safe position of journeyman to another, and masters were often glad to have the resource of several skilled pairs of hands to assist them in their work. It was often the case that it was more prestigious to work as a journeyman for a well-known and respected master than to struggle to set up a workshop of one's own.

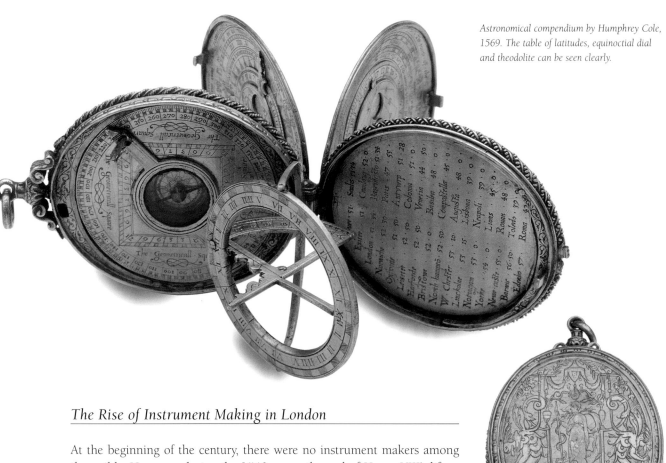

Astronomical compendium by Humphrey Cole, 1569. The table of latitudes, equinoctial dial and theodolite can be seen clearly.

The Rise of Instrument Making in London

At the beginning of the century, there were no instrument makers among the guilds. However, during the 1540s, near the end of Henry VIII's life, a few foreign makers came and settled in London. The earliest known maker was a man called Thomas Geminus, who came from the Netherlands, and may have been driven out by religious persecution. These emigrés probably trained young Englishmen in the craft of instrument making and a small group of indigenous makers emerged during the second half of the century. However, instrument making was not as lucrative a craft as it appears to have been in the mainland of Europe – fewer people were interested in

Exterior, showing Juno with her peacock.

Interior showing the perpetual calendar and the lunar volvelle.

Interior showing the table and instrument used together for calculating the time of high tide.

buying instruments and the makers were forced to supplement their income by other occupations. Some of them also worked as map-engravers and surveyors; others were booksellers; the chief artisan of the period, Humphrey Cole, had a job in the Royal Mint.

Humphrey Cole was born in the 1530s in the North of England, moving down to London and probably gaining membership of the Goldsmith's Company. In 1564 he was listed as an assistant at the Mint, and four years later he set up a workshop 'Near the North Door of St Paul's', as he described it. Apart from his work at the Mint, he was an engraver, a die-sinker and an expert in mining and metallurgy. However, he is remembered for the fine quality of his instruments which left a lasting legacy after his death in 1591, and which included alidades, armillary spheres, astrolabes, astronomical compendia, gunner's callipers, logs, nocturnals, quadrants, rules, sundials and theodolites.

One of Cole's finest instruments is the astronomical compendium which is held at the National Maritime Museum. It was long known as 'Drake's astrolabe', completely erroneously since it is not an astrolabe and there is no known connection whatsoever with Sir Francis Drake. However, it is clear that it was made for a very wealthy patron and it had a long history in the hands of the nobility: its first known owners were the Stanhope family, through whom it eventually came into the possession of King William IV. He in turn gave it to the Greenwich Naval Hospital in 1833 where it remained until it was transferred to the Maritime Museum in the 1930s.

The compendium is inscribed *Humfray. Colle . made this . diall . anno . 1569.* So it is an early instrument for Cole, produced only shortly after he opened his workshop near St Paul's. It is a particularly complex instrument, carrying a lunar volvelle, a perpetual calendar, a table of latitudes, a universal equinoctial dial, a compass, a simple theodolite, a table of ports and harbours, and an instrument for calculating the time of high tide for all the ports listed in the table. The outside of the compendium is decorated with portraits of Jupiter and Juno, which Cole copied from a series of images of classical gods and goddesses, made by Etienne Delaune in France in the 1560s. Jupiter is shown with his customary thunderbolt, eagle and sceptre while Juno appears with her usual attribute, the peacock.

Inside the lid of the compendium the first of a plethora of instruments appear. These are a lunar volvelle and a perpetual calendar. The former is unremarkable, except perhaps for the inclusion of a scale marked with the signs of the zodiac and the title: '*The course of the sonne and Mone throughe 12 Signes*'. However, the perpetual calendar is more interesting. It is titled '*A kalender with ye Saint^S daies and moueable feastes for euer.*' The saints' days listed include the major fixed feasts (Christmas, Epiphany, the Circumcision of Christ, the Annunciation, etc.) as well as the most important saints – the apostles, Mary Magdalene, John the Baptist, All Saints, All Souls, the Conversion of St Paul – and some minor inclusions such as St George, the patron saint of England. Otherwise the fixed calendar lists the dates for the beginning of each sign of the zodiac, according to the Julian Calendar (the Gregorian Calendar had not yet been devised). The calendar is also important for its omissions: it does not include the Assumption of the

Virgin, celebrated in Catholic countries on 15 August. England was by this point a Protestant country, and some of the Catholic festivals had already been removed from the Church's calendar. Within the circles of the calendar is a table for calculating the date of Easter, giving the possible dates of Easter with the corresponding dominical letter and prime. Within this table is a second table giving the primes and epacts which refer to the dates of the full moons through the year, and which are also necessary for calculating the date of Easter. The method of calculation is very complicated, but it can be found at the beginning of most versions of the Book of Common Prayer.

Turning over the leaf carrying the perpetual calendar we find a table of latitudes which concentrates, as one might expect, on English towns (including the slightly unusual appearance of Basingstoke), but it does also include major European cities. On the opposite leaf is a compass set into the centre of a simple theodolite. The inclusion of the theodolite is unusual and it transforms the compendium into a surveying as well as an astronomical instrument. The theodolite had only recently been invented; this is one of the earliest examples of the instrument still extant. It differs slightly from what we would now term a theodolite because it will only take readings in a horizontal plane and not in a vertical one.

Between the latitude table and the theodolite is the dial. This is the only dial to appear on this particular compendium and it is in the form of an equinoctial dial. It seems rather different from the equinoctial dials which can be found on the outside of diptych instruments. This is because the hour plate is pierced and the gnomon is a long rod passing through the centre of the circle. The quadrant which is visible behind the gnomon is a latitude quadrant, marked with degrees from zero to ninety. It passes through a groove in the semicircle supporting the hour plate and the point at which it crosses the semicircle marks the latitude for which the dial is set.

The final section of the compendium carries a table of the principal ports and harbours around the coast of Britain and the west coast of Europe, and an *INSTRUMEN TO KNOWE THE EBBES AND FLVDDES*. The table and the instrument are used in conjunction for calculating the time of high tide at a particular port. With the inclusion of this table and tide calculator the object is now transformed into a navigational instrument, as well as one for surveying and astronomical uses.

Cole's compendium is a really remarkable instrument. It is a true compendium with its inclusion of so many different instruments and tables. It is accurately made and its sundial would tell the time to the nearest quarter of an hour (quite sufficient for the period). Moreover, the quality of engraving on the object is exquisite: not only the vignettes of Juno and Jupiter, but also the execution of the flourishes on the other sections mark the compendium as the work of a master craftsman at the height of his powers.

This final example has shown the fine workmanship of one of the earliest of the English instrument makers. In the next chapter we will discover how the English instrument-making trade (including the making of sundials) blossomed in the seventeenth century, and began to rival that of the continental makers.

Chapter Five

ENGLISH ACCENTS

The Cole compendium mentioned in the previous chapter was one of the early signs of an upsurge of sundial production in England. Cole was one of the early instrument makers in London, and after him a slow but steady trickle of craftsmen produced scientific instruments through the latter part of the sixteenth century. It is difficult to know whether any of these makers were trained by Cole, but we do know that lines of succession in the trade can be traced from the 1580s, when Charles Whitwell was apprenticed to Augustin Ryther, and later became the master of Elias Allen, one of the greatest makers of the seventeenth century. Compendia similar to that made by Cole were produced by both Whitwell and Allen; the quality was comparable to that of the Cole compendium and two of Allen's instruments carry the royal coat of arms, denoting that they were made for the king, either James I or Charles I.

A New Trade in London

The creation of an instrument-making trade in London led to the establishment of workshops for the manufacture of instruments and of the apprenticeship method of training for young makers. Because the trade was so undeveloped, there was no central guild to which the instrument makers were attached, and consequently they are found as members of guilds as varied as the broderers (embroiderers), the goldsmiths, the joiners and the grocers. The number of instrument makers was particularly large in the Grocers' Company, and this remained a popular guild even after the establishment of the Clockmakers' Company in 1632. The Clockmakers made several attempts to gather in the instrument makers and to enforce more rigid policing of the trade. This had been lax before and had allowed for a substantial degree of variation in the quality of instruments and of the discipline in the makers' workshops, especially with regard to the number of apprentices permitted. However, the efforts of the Clockmakers' Company produced little effect: makers joined the new company if they felt that it would be useful to them; otherwise they took very little notice of attempts at coercion. This independence of the instrument makers no doubt enabled a greater freedom in their relationships with their customers and the instrument designers, and a greater opportunity to play the market.

The Rise of English Mathematics

Why was it that a good market existed in England for scientific instruments (including sundials) at the very end of the sixteenth century and not before? And what led to makers of the mid-seventeenth century being able to support themselves solely from the sale of their instruments, unlike Cole and his contemporaries? The reason was the increase in interest in mathematics during the latter part of Elizabeth's reign, and more obviously during the following century. There were various reasons for this upsurge, the most important being the introduction of textbooks in the vernacular, the influence of John Dee and the creation of Gresham College.

Until the middle of the sixteenth century almost all academic books were in Latin, the lingua franca of the university system. However, 1542 saw the publication of the first English arithmetic, *The Grounde of Arts*, by the physician Robert Recorde. This was the first in a rapidly expanding wave of English texts in the mathematical sciences which grew through the sixteenth century and really blossomed during the seventeenth. These books opened the way for the non-university-trained to learn something of the mathematical sciences (which included astronomy, navigation and surveying as well as geometry, arithmetic and trigonometry).

Engraving of John Dee, artist unknown.

Secondly, an important role was played by the Elizabethan magus and polymath, John Dee. Dee had travelled widely in Europe during his youth and had gained a love for mathematics which was the reason for his encouragement of its use. He believed that there were good political reasons for England to develop mathematical skills, firstly because such techniques were important for military engineering, but more significantly because knowledge of mathematics and the use of mathematical instruments were essential for oceanic navigation and therefore the establishment of an overseas empire. Dee had some success in promoting his cause, at least to the extent of a mathematical lectureship being established in London in 1588, expressly for the citizens and not for academics. However, the lectures lapsed in the mid-1590s and were not reinstated, but a substitute came with the creation in 1597 of Gresham College.

Sir Thomas Gresham was economic adviser to Queen Elizabeth and founder of the Royal Exchange. He believed that the economy of the City of London would be aided by the education of at least the merchant class of the city. To this end he left directions in his will for the establishment of a college which would provide adult education in various subjects. There were to be seven professors of this college who would lecture on the subjects of Rhetoric, Law, Divinity, Medicine, Music, Geometry and Astronomy. They were to be paid £50 a year and their task was to give two lectures a week in their subjects in both Latin and English. The lectures were to be open to any who wished to attend.

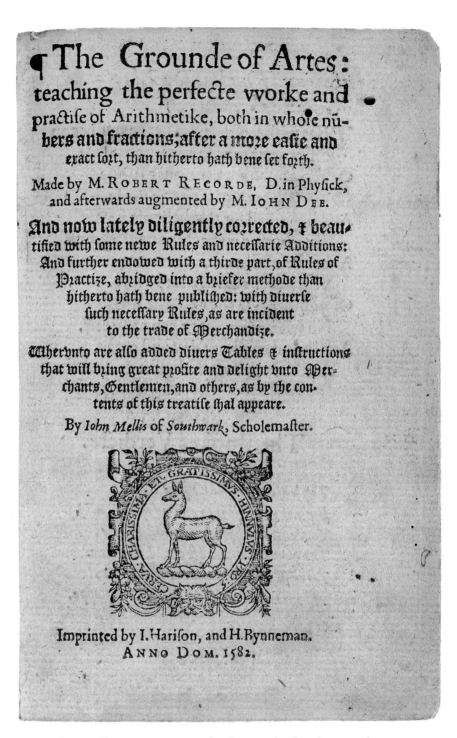

Title page of The Grounde of Artes *by Robert Recorde, 1542; this edition 1582.*

Gresham College was important for the growth of mathematics because it was the first place in England where chairs in mathematical subjects were established. Not only that, but the professors were required to lecture on the practical applications of their subjects: the geometry professor gave courses on surveying while the astronomy professor taught navigation and the use of mathematical instruments. The College still exists today, with the chairs being held now for a fixed term rather than for life. While the College

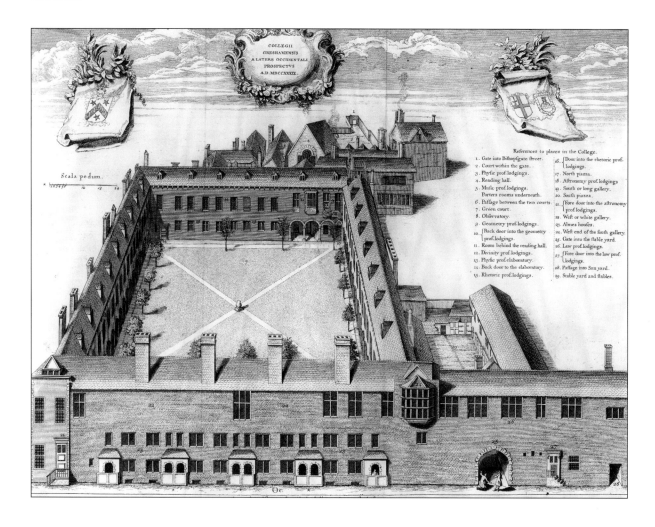

Engraving of Gresham College, taken from John Ward's Lives of the Professors of Gresham College, 1740.

was not an unqualified success, in its early years several good mathematicians were appointed to the chairs of geometry and astronomy and through them the building became a meeting point for mathematicians and their hangers-on. Meanwhile the general increase in interest in mathematics led to the establishment of numerous private teachers throughout the capital.

The first professor of geometry was Henry Briggs, a graduate of Cambridge and a gifted mathematician. He became the focus for the gathering of a group of mathematicians and natural philosophers (the equivalent of the modern scientist). These included two people who are particularly important to our story: Edmund Gunter and William Oughtred.

Edmund Gunter

Edmund Gunter was born in Hertfordshire in 1581 and passed his undergraduate years at Oxford between 1600 and 1603. During this time his interest in mathematics (which had already shown itself in a fascination with sundials) flourished and he began to design his own mathematical

Title page of De Sectore et Radio, *1624.*

instruments. In about 1606 he wrote a Latin manuscript describing a 'sector', a calculational instrument which he had developed from an earlier surveying instrument. However, he did not publish this work until 1623, when it appeared in English together with the description of various other instruments under the title *De Sectore et Radio or The Description and Use of the Sector, the Crosse-staffe and other instruments*. In 1615 Gunter took holy orders and was appointed rector of both St Mary Magdalen, Oxford and St George's, Southwark. He began to spend time in Gresham College where his friend Henry Briggs was professor of geometry. Gunter became the third Gresham professor of astronomy in 1620 and remained at Gresham until his death in 1636.

During his time at the College, Gunter devoted most of his energy to the development of instruments and to the investigation of the newly discovered logarithms. He devised the first table of logarithms of trigonometrical functions, dubbing them 'artificial sines and tangents'. These were published as *Canon Triangulorum; or, Table of Artificial Sines and Tangents* (1620) in which Gunter also introduced the terms 'cosine' and 'cotangent'. Gunter lectured on much of his logarithmic work at Gresham, as well as teaching the use of mathematical instruments in astronomy and navigation. He was a firm advocate of the use of instruments in mathematics for easing the work of various mathematical practitioners, notably surveyors and navigators, and his instruments were designed with these aims in mind.

Gunter's New Quadrant

I have mentioned Gunter's work on the sector and it was in this book that he made his major contribution to the development of sundials. Among the 'other instruments' of which he speaks in the title was a quadrant of his own devising, which could be used for various purposes in astronomy and surveying, but whose chief function was time-telling. It became the last important innovation in altitude dials, whose popularity had waned with the introduction of portable horizontal and equinoctial dials.

As I said in chapter two, the quadrant had been used before as a time-telling instrument in the form of the *quadrans vetus*. However, a second form of quadrant was designed in 1288 by the Arabic scholar Jacob ben Machir ibn Tibbon (1236–1305). This quadrant, based on an astrolabe plate, worked on a system of equal hours (rather than the unequal hour system favoured for the *quadrans vetus*) and was used only for one latitude. A relatively simple version of the new quadrant was produced by Thomas Fale at the end of the sixteenth century and described in his book, *Horologiographia: the Art of Dialling*, published in 1593. It was not an independent instrument since it required the use of tables in order to find the place of the sun in the zodiac for a particular date, and thus to set the instrument for time-telling. It was not until Gunter developed his quadrant

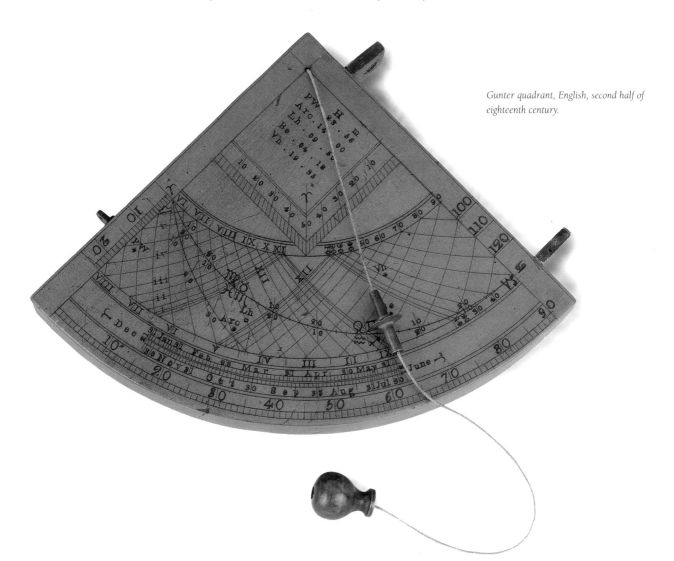

Gunter quadrant, English, second half of eighteenth century.

Another unsigned quadrant from the same period, showing the nocturnal on the reverse side.

Nocturnal, English, seventeenth century.

from this instrument that a new dial was created which could be used without reference to any other source of information.

The Gunter quadrant as described by its inventor carried the following instruments: a set of hour lines and a date scale for use in time-telling; a set of lines for calculating the azimuth (direction East or West) of the sun; a shadow square for use in surveying; and, on the reverse, a nocturnal for telling the time at night. I will explain the use of the quadrant for time-telling, and then look briefly at its other functions.

The picture on the previous page shows the front of a typical Gunter quadrant. The parts which are required for telling the time are the degree scale running along the curve of the quadrant, the date scale immediately above it, the curving lines on the left which are marked with the hours in Roman numerals, and the sights on the right-hand edge of the instrument. There is also a string attached to the hole at the apex of the quadrant, carrying a plumb bob and a bead, similar to those which appeared on the *quadrans vetus*, Regiomontanus dial and *navicula*. As with those instruments the first preparation is to set the bead correctly, though here it is only for the date and not for the latitude (the Gunter quadrant can only be used in the latitude for which it was made). The string is placed taut across the correct date and the bead is moved along the string until it rests on the line for midday (the twelve line). After this the quadrant is raised vertically and tilted until the sun's rays fall through the pinhole of the top sight on to the pinhole in the bottom sight, with the string being allowed to hang freely. The string is then clamped against the degree arc to make sure that it is not moved again. When the quadrant is lowered the position of the bead in the hour lines will show the time of the day. As with the similar dials we met in chapter two the procedure for telling the time is complicated and very different from that used for the direction sundials we have been discussing in the last two chapters. However, it works well, giving good results, and the quadrant had other functions which made it a useful addition to the sundial family.

The other lines on the front of the quadrant can be used for finding the time at which the sun rises and sets, for finding the sun's position in the zodiac at a particular time of year, for finding the direction of the sun at a particular time of day, and for performing various other astronomical functions. The lines in the upper part of the quadrant are part of a 'shadow square' and are used in surveying for measuring the heights of landmarks, just like the shadow square on the *navicula*.

The reverse of the quadrant almost always carried a nocturnal as shown on page 73. The nocturnal was an instrument for telling the time at night by means of the stars. It was developed in the sixteenth century and was most often found as a separate instrument but was sometimes included on astronomical compendia and quadrants (such as the one shown here). The nocturnal depended on the fact that, just as the sun changes its position depending on the time of day and the season of the year, so the stars alter their position in the sky. If it is known where a star or group of stars should be with respect to the Pole Star on any one night, then the time of night can be found. The nocturnal reduces all the information required into a very simple instrument. Here we see that five of the circumpolar constellations have been engraved in their correct positions around the Pole Star: the Great Bear (or Wagon, Plough, Saucepan or Charles' Wain), the Little Bear, Cassiopeia, Cepheus and the Dragon. Around them a date scale is marked and the whole disc on which the constellations and date scale are engraved can rotate within a scale carrying the hours of the day and night. The principle of the nocturnal's use is this: the user observes which stars are near the meridian (the North-South line, with North being marked by the Pole Star) and rotates the disc of the nocturnal until the position of the stars in the sky is copied on the instrument (the twelve o'clock line being the meridian line). The current date will now be positioned against the time of night.

An Amateur Quadrant

The example which I have picked out for closer study is not an entirely typical example of a Gunter quadrant. While the front is fairly recognisable as a Gunter quadrant it does not include a date scale anywhere and the nocturnal on the back has been replaced by a spiral scale. You might think from a glance at the poor quality of the letters and numerals on the instrument that this was not a genuine instrument at all but a fake. However, this is not

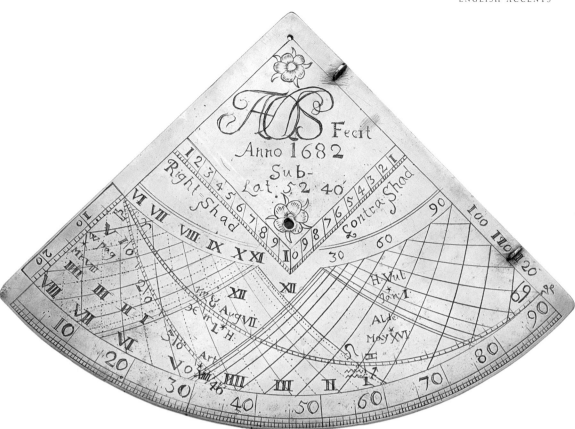

Gunter quadrant by unknown amateur, 1682.

the case, and it appears that this quadrant is a good example of the kind of instrument that was occasionally produced by amateurs during the seventeenth century. While sundials were normally made by professional makers, examples by amateurs do exist and the Gunter quadrant appears to have been a relatively popular instrument for amateurs to attempt to make. The production of the dial would demonstrate the skill of the maker, who would require a reasonably strong mathematical background to understand the principles involved. In particular, a solid grounding in trigonometry was needed in order to mark the curves of the hour lines and other important curves on the instruments. However, Gunter gave clear instructions in his book for making the quadrant and these could have been followed fairly easily by anyone who had a good grounding in mathematics.

Looking more closely at this quadrant we return to the puzzling fact that it has no date scale. It would seem that this would prevent the instrument from being used for telling the time. However, there are other means for adjusting the bead: that given above was the most simple method for finding the sun's position relative to the zodiacal belt (known as the ecliptic) or the celestial equator, which is the purpose of positioning the bead on its string. Both the other means are available on this instrument. The scale down the left-hand side is a declination scale and marks how far the sun is above or below the celestial equator. If the string is set against the left-hand side of the quadrant, and the declination of the sun is known, the bead can be moved to the correct point on the scale and will then be correctly adjusted. Alternatively, the ecliptic line can be used: this is the line which

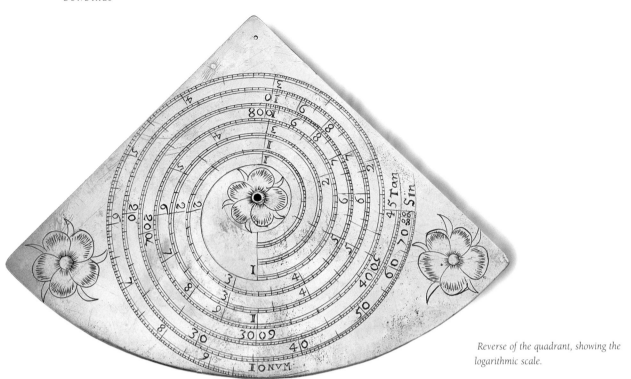

Reverse of the quadrant, showing the logarithmic scale.

runs in an arc from halfway down the left side of the quadrant to the bottom of the right side. It marks the date in terms of the signs of the zodiac and the string can simply be stretched across at the right position and the bead moved until it is on the line of the ecliptic; the bead is then correctly calibrated. So we can see that this particular amateur maker was sufficiently confident of his astronomical knowledge to dispense with the simple date scale and to use other means for setting up his instrument. Similarly, the maker clearly felt no need for a nocturnal on the reverse and has instead substituted a spiral logarithmic scale for making calculations involving the trigonometric functions of sine and tangent.

What can we discover about this mysterious person? We know that the instrument was made in 1682, for a latitude of 52° 40′ N, by someone whose monogram appears in the centre of the shadow square. Unfortunately the monogram is not clear enough to be decipherable, although it obviously ends in an 'S'. The latitude is fairly accurate for Norwich, Peterborough or Leicester, so all of these are likely candidates for the place of origin. The maker was clearly well-educated in mathematics and so probably had a university training. Cambridge might be the more likely university, since it is closer to the towns which I have mentioned as possible home towns for the maker, and it was usual in the seventeenth century for men to go to the nearest university. However, that is really all we can tell about the maker of this quadrant.

One thing that this quadrant does show is an odd relapse to the days when the makers of instruments were also the users of them, when mathematicians and astronomers were the main people concerned in producing instruments and the craft of instrument maker had not been established.

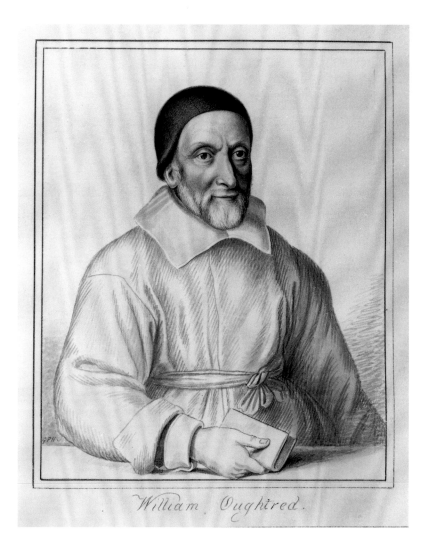

William Oughtred.

*William Oughtred, copy by George Perfect
Harding of an original engraving by
Wenceslaus Hollar.*

William Oughtred and his Designs

William Oughtred was a rather different man from Edmund Gunter. He was
born at Eton on 5 March 1574, the son of the school's registrar. In 1592 he
was admitted to King's College, Cambridge, where he took both his BA and
his MA and was a Fellow from 1595 until 1603, mainly concentrating on
research in mathematics. Having been ordained he spent five years as vicar
of Shalford in Surrey, where he married a woman with the unusual name of
Christsgift in 1606, before moving to become rector of Aldbury in Surrey in
1610, where he remained until his death in 1660. Early in his time at
Aldbury he was fortunate enough to attract the notice of the Earl of
Arundel, who had a residence in Oughtred's parish. Arundel appointed
Oughtred as tutor to his son, Lord Howard, and, as a result, from the late
1620s, Oughtred began to spend significant periods of time at Arundel
House in the Strand. This base in London allowed him to have contact with
the circle of mathematicians who were linked to Gresham College, mostly
through their acquaintance with Henry Briggs, whom Oughtred himself

had known for many years. It was also an opportunity for the mathematician to develop an earlier association with the instrument maker, Elias Allen, whose workshop was only a few minutes' walk from the Earl's house.

Oughtred is best known for his book, *Clavis Mathematicae* (*The Key to Mathematics*), which was first published in 1631 (under an earlier, less well-known, title) and which was an early algebra textbook. Various scholars visited him both in London and at his home in Aldbury to learn from him and to be instructed in algebra. However, Oughtred was also interested in the design of instruments which he had first taken up while at Cambridge. It was during his university years that he devised two important instruments for the history of sundials, but he did not publish a description of them until much later in his life. Later, having heard of the discovery of logarithms, he produced the first forms of slide rule (in circular form and later in the common straight form used until the 1970s).

Oughtred's Instruments

In 1636, Oughtred published *The Description and Use of the Double Horizontall Dyall*. This object was a horizontal dial which carried both a standard horizontal dial scale (for use with the normal sort of gnomon) and a second set of hour lines which could be used with the vertical edge of the gnomon set at the centre of the dial plate which could also demonstrate various astronomical motions. It need not concern us here (since it never appeared in a portable form) more than to note that the surviving examples show the close connection between the author and Elias Allen, who made many of these instruments, and who also passed the method of production

Double horizontal sundial by Elias Allen, 1630s.

Universal equinoctial ring dial by Elias Allen, 1630s.

on to his apprentices in the trade. However, the second edition of the book, which appeared in 1652, is of more interest for the history of portable dials, because it included an appendix: 'The Description of the General Horological Ring'.

This 'ring' was, in fact, a sundial which Oughtred had invented much earlier in his life, probably while he was at Cambridge. It was developed from the astronomical ring, which was first described by Gemma Frisius and others, early in the sixteenth century. That in turn had been a development of the armillary sphere, a device which had been known since the time of Ptolemy. The armillary sphere was an astronomical instrument for demonstrating the important features of the heavens and the various celestial circles such as the equator, the ecliptic and the tropics. The instrument known as the astronomical ring, which actually consisted of three rings, was a general instrument for making astronomical observations, but it could also be used to find the time if necessary. However, when Oughtred developed his horological ring he created an instrument which was only intended for time-telling.

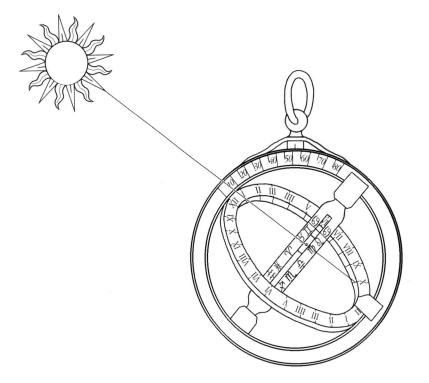

A universal equinoctial ring dial in use.

Oughtred's dial (now known as the universal equinoctial ring dial, for reasons which will become apparent) was perhaps the most basic model of the heavens which could be devised. It consisted of two rings and a bridging bar inside the inner ring. The outer ring was known as the meridian ring, and represented a great circle passing through the North and South celestial poles. Within this ring and set at right angles to it when the instrument was opened was the equinoctial ring which represented the celestial equator. The bridging bar was set between the poles of the meridian ring and so represented the axis of the world, like the gnomon of a horizontal sundial. Within this very simple model of the heavens the passage of the sun could be traced and used to find the time.

The use of the universal equinoctial ring dial is best shown by a diagram. The meridian ring carries a latitude scale and the suspension ring is moved around the latitude scale until it is set at the observer's latitude. The pinhole in the slider set between the bars of the bridge is moved until it corresponds with the correct date as marked on the date scale on the bridge. Then the instrument is opened out until the meridian and equinoctial rings (the latter of which is marked with an hour scale) are at right angles, and the bridge is adjusted so that the pinhole is turned towards the time as roughly estimated by the user. Finally the dial is held up and rotated until the beams of the sun pass through the pinhole and fall on the hour scale, indicating the time.

This sort of dial does not have a compass, but that does not mean that it is an altitude dial, dependent on the sun's height in the sky to give the time. It is actually a direction dial, like the horizontal, vertical and equinoctial dials which we have considered in previous chapters. It does not need

Universal equinoctial ring dial by Elias Allen, 1630s.

a compass because when it is oriented so that the sun passes through the pinhole and falls on the hour scale it is automatically set in the correct position and the meridian ring will indeed be lying in a North-South plane, parallel with the great circle which it represents, while the equinoctial ring will be parallel to the equator. This means that this dial is both clear and accurate because it does not create the confusions over morning or afternoon hours arising from altitude dials, nor does it have to be adjusted for magnetic variation (a problem with compasses which will be discussed in the next chapter, but which had already hindered the accuracy of earlier dials containing compasses). The only problem with this type of dial is that it cannot be used close to midday, since at that time the shadow of the meridian ring obscures the pinhole and prevents the sun's rays from falling on the equinoctial ring.

Later versions of the universal equinoctial ring dial carried an altitude quadrant on the reverse for measuring the height of the sun (rendering the instrument particularly useful to sailors who needed this measurement to calculate their latitude). However, in the form devised by Oughtred, it was only the time-telling attribute that played any part.

ELIAS ALLEN.
Apud Anglos Cantianus iuxta **Cunnbridge** natus, Mathematicis Instrumentis are incidendis sui temporis Artifex ingeniosissimus.

Obijt Londini, prope finem Mensis Martij, Anno a Christo nato 1653, suæque ætatis.

Elias Allen, engraving by Wenceslaus Hollar, 1666, after a painting by Hendrik van der Borcht, c.1640.

A Dial by Elias Allen

This particular universal equinoctial ring dial (see illustrations on p.79 and p.81) was made by Elias Allen, the maker with whom Oughtred had close connections. When the account of the dial was published in 1652, the only maker advertised was Allen, although he was over sixty by this time, and, in fact, died the following year. It is possible that he was the only maker to produce the dials before the publication of the book, with the exception perhaps of the apprentices who trained under him. The earliest dated dial of this type was made by Anthony Thompson in 1652, but Allen must have been making them long before that date.

The dial itself is a very simple example of its kind. It is small – much smaller than the ring dials that were manufactured during the eighteenth

The 'Madagascar Portrait' of the Earl and Countess of Arundel by Anthony Van Dyck, 1639.

century – and therefore perhaps less accurate than later versions, but certainly more portable (some later English dials of this type reached a diameter of ten inches). The Allen dial carries only a latitude scale, an hour scale, a date scale on the bridge (later versions added zodiac scales and scales of solar declination) and a table of latitudes covering most of the remaining available space of the dial. This table of latitudes may give us some indication about the original owner of the instrument. It is possible that the town listed immediately to the right of the latitude quadrant indicated the place of manufacture or where it was most likely to be used. In this case the town is Nottingham, clearly not the place of manufacture, and so perhaps the home town of the owner. Another universal equinoctial ring dial by Allen in the Science Museum carries this same 'base latitude' and may have been made for the same customer. The other towns listed on the dial include major places in Britain and some of the coastal cities of the Low Countries, France and the Iberian Peninsula, along with Paris and Madrid. It seems then, that the original owner might have had cause for travel to these ports; he may well have been a merchant.

The dial is not dated, but we have clues from other sources to suggest an approximate time at which it might have been made. One of these sources is a letter; the other two are portraits. The letter was written by the botanist, John Beale, to the natural philosopher, Samuel Hartlib, in 1657, and Beale commented that 'above 20. yeares agoe, I was with Elias Allen over against Clements Church, whilst hee made the Ring-Diall, universall for all Climats.' So the ring dial must have been available from at least as early as 1637. The first portrait is that of Allen himself, made at some point around 1640, and which shows the maker seated at his workbench, with a ring dial prominent in the foreground of the picture. However, the most certain dating comes from Van Dyck's portrait of the Earl and Countess of Arundel painted in 1639 to mark the Earl's plans for an expedition to Madagascar (which were never realised) showing Lady Howard holding a universal equinoctial ring dial. From this evidence we can assume that the present example may well have been manufactured during the 1630s, twenty years before the first dated example appeared.

The Gunter quadrant and the universal equinoctial ring dial in particular remained common instruments throughout the seventeenth century and during the eighteenth century the ring dial was the most popular English design to be produced. It was manufactured elsewhere in Europe to a certain extent, but there other designs were more prevalent, as we shall see in the following chapter.

Chapter Six

A CONTINENTAL SELECTION

Introduction

While the universal equinoctial ring dial was produced by some European makers, the vast majority of this form of dial were manufactured in England, continental makers preferring different types. During the latter part of the seventeenth century, three different forms of sundial were developed in continental Europe. These dials were characteristically manufactured in one town, in the same way that ivory diptych dials had been peculiar to Nuremberg in the latter part of the sixteenth and the first half of the seventeenth centuries. This concentration in one city had been unusual, at a time when each type of dial (pillar dial, ring dial, quadrant, horizontal dial, etc.) was made in many different places. However, during the seventeenth century, regional manufacture of particular forms of dial became much more prevalent, and associations between a form and a town or small area continued until the decline of the sundial in the nineteenth century.

Magnetic azimuth dial by Charles Bloud, c.1660.

It is unclear why such patterns of production developed. Perhaps makers from one town were adept at marketing their particular type of dial and so created a fashion in that place for possessing that form. Be that as it may, we find that three continental centres of dial-making arise in the seventeenth century – Dieppe, Augsburg and Paris – and we shall look at these in turn. Their associated dials were the Dieppe magnetic azimuth dial, the Augsburg dial and the Butterfield dial (which was made in various towns in France, but its production centred on the capital).

Augsburg dial by Johann Georg Vogler, mid-eighteenth century.

Butterfield dials by Michael Butterfield, c.1700.

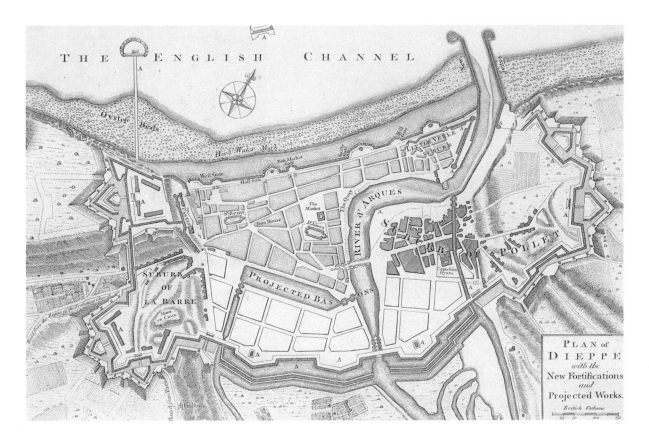

Plan of Dieppe, artist unknown, 1694.

Dieppe

Dieppe is a seaport in Normandy, about fifty miles south of Boulogne and Calais. There had been a settlement at the mouth of the river Béthune since Roman times, but the name 'Dieppe' was probably not given to it until about 1000 AD. As part of Normandy the town came into the hands of the English following the Norman Conquest in 1066, and remained an English town for most of the years until 1204. The trade with England led to Dieppe's growth as the most important port in Normandy and, although it was reduced to rubble during the Anglo-French wars of the 1190s, it soon rose to prominence once again. Flourishing overseas trade increased the wealth of the town, particularly during the time of Jean Ango. Ango was governor of Dieppe in the first half of the sixteenth century and his escapades against the Portuguese treasure fleets brought great prosperity to his home town.

By the mid-seventeenth century Dieppe was one of the most important remaining Huguenot communities in France. Huguenot was the name given to the French Protestants and probably derives from the German *eidgenoss* (oath-fellow) which was used to refer to the groups of Swiss Protestants who fought beside their French co-religionists during the earlier Huguenot wars of the sixteenth century. These wars, which lasted throughout the sixteenth century, and which were to continue to have repercussions throughout the following century, were some of the bloodiest civil wars ever fought in Europe. It was estimated that by the time peace came with the accession of Henry IV in 1594, two million people had been killed in the previous seventy years. Many French towns lay in ruins and the land was so devastated that it could not support the remaining population.

However, the Edict of Nantes (declared in 1598) brought some degree of peace between Catholics and Protestants. The Edict permitted considerable (though not complete) freedom of worship to Huguenots throughout the country: they were declared eligible for public offices, for entrance into schools and for admission to hospital when sick. While Catholicism remained the state religion and dominated the political processes of the seventeenth century, most Huguenots were left in peace for many years. The exception was the siege of La Rochelle in 1628, in which 20,000 people starved to death rather than submit to Cardinal Richelieu. However, this siege was largely political rather than religious in motivation: La Rochelle was so strongly defended that it always provided a means for foreign Protestant powers to make attacks on France in relative safety, with the assistance of the townspeople. With La Rochelle stripped of its defences, the threat of invasion was removed.

Remaining Huguenots kept to particular towns in France and the ports of Normandy were especially popular. Among the citizens of Dieppe were a group of Protestant instrument makers who were responsible for the popularisation of the magnetic azimuth dial.

The Origins of the Magnetic Azimuth Dial

Azimuth dials had been in use for some time when the first magnetic azimuth dial was made in Dieppe early in the 1650s. An azimuth dial tells time by reference to the local direction of the sun – how far East or West it is along the horizon – rather than the direction relative to the axis of the Earth. As we saw in chapter three, there is no constant correlation between the local direction of the sun and the time of day. This was the reason why gnomons on horizontal dials had to be inclined at an angle parallel with the Earth's axis. However, it was realised at some point during the Renaissance or early modern period, that a vertical gnomon could be used to tell the time by the direction of its shadow, if two provisos were observed. The first was that the hours must be marked out around an elliptical scale rather than a circular one, with each hour indicated by a point rather than a line. The second was that the gnomon itself must be able to move along the short axis of the ellipse on a scale graduated for the date. If such adjustments were made, the dial would tell the time accurately throughout the year.

Such dials were constructed as fixed dials in various locations during the

Analemmatic dial by Thomas Tuttell, c.1697.

sixteenth century. In the succeeding century it was realised that they could become useful as portable dials as long as they were combined with a standard horizontal dial. While neither of these dials could be oriented correctly on their own without the aid of a compass, the issue of orientation became trivial when the two were combined. Since they operated using different principles, if they were oriented until they were in the sole position when they were both telling the same time, this must be the actual time of day.

These combined dials (known as analemmatic dials) were usually rather large and cumbersome. However, at the same time as they were being developed a rather smaller form of azimuth dial was invented. The invention was claimed by Charles Bloud of Dieppe, and it is certainly true that this particular form of azimuth dial was solely associated with Dieppe during the years in which it was produced.

The Dieppe Magnetic Azimuth or Bloud-type Dial

Dieppe magnetic azimuth dial by Jacques Senecal, c.1660. The inside of this dial has a lunar volvelle in the upper leaf and the main dial set into the lower one.

The dial appears in the form of a diptych, and was commonly made of ivory, so resembling the Nuremberg diptych dials. The outside of the upper leaf was often engraved with an equinoctial dial (in the same manner as the Nuremberg dials) or a polar dial (a dial on which the hour lines lie parallel to each other, but otherwise tells the time much as an equinoctial dial would, with the aid of a pin gnomon set at the centre of the hour lines). The inner surface of this leaf often carries a lunar volvelle, (as mentioned in earlier chapters): the nature of the dial, as we will see below, allowed its use at night during the weeks either side of the full moon.

The main part of the dial occupies the lower leaf of the diptych. Here we find the typical recessed bowl for the compass, but rather than being marked solely with the directions, this compass is also equipped with an elliptical ring on which the hours of the day are marked in an anticlockwise direction. The reason for this is that as the dial is turned in a clockwise direction following the path of the sun, the magnetic needle is stationary and so appears to be moving anticlockwise over the hour scale. An ordinary horizontal dial with a string gnomon is sometimes found marked around the edge of the compass bowl as an additional feature.

The gnomon for an azimuth dial needs to be adjusted for the date; in a magnetic azimuth dial this process is actually carried out by moving the hour scale while the gnomon (the compass needle) stays where it is. It can be seen from the illustration that there is a groove on the base of the compass bowl, along which the hour scale can be moved. The movement of the scale is controlled by a rotatable disc on the underside of the leaf. The disc is divided into the months and days of the year, and is turned until the current date is lined up against a hand on the ivory of the leaf, at which point the hour scale is correctly set.

These magnetic azimuth dials are extremely simple to use. Once the dial

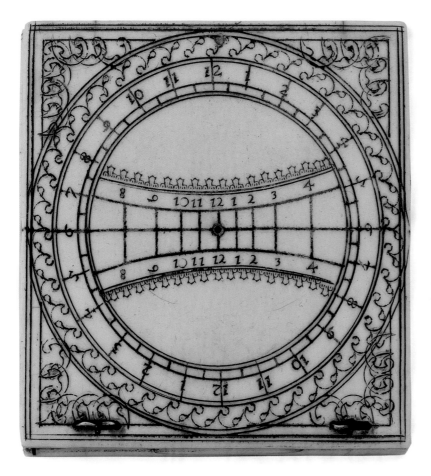

Dieppe magnetic azimuth dial, unsigned, c.1660. This plate carries both an equinoctial dial and a polar dial.

has been set for the date, the upper leaf is opened and the dial is turned towards the sun, until the shadow of the upper leaf completely covers the lower leaf. At this point the magnetic needle will show the correct time on the hour scale.

Most of the magnetic azimuth dials carry a table of latitudes in the base of the compass bowl. This is sometimes enlarged to form a gazetteer giving details of the attributes of various towns (always French). The dial illustrated on the previous page could tell you that Rouen is the seat of an archbishop (ar), a provost-marshal (pr), a bailiff (bai) and a seneschal (sen); that it possesses a university (vn), and a port (por) and that it is a duchy (du), the capital of a treasury subdivision (gen for *géneralité*), the centre of an electoral district (élé) and a town acting as a grain supply for a whole region (gre for *grenier*); 'mo' probably stands for 'moulin' (flour mill) but this is uncertain.

The magnetic azimuth dial had one major advantage over other forms of direction dial: it could be used when the sun was not shining brightly enough to cast a strong shadow. The orientation of the dial depends on it being turned towards the sun; it is possible to find the direction of the sun when it is only shining weakly, or is behind thin cloud. Such weak sunlight would be insufficient to cast a good shadow of the gnomon on an ordinary

Dieppe magnetic azimuth dial by Charles Bloud, c.1660. The volvelle on this leaf is used for setting the dial for the correct date.

direction dial, but this dial tells the time by means of the compass needle, and so strong shadows are unnecessary. This is the reason why it could also be used in moonlight.

However, it had one major disadvantage which it shared with all forms of magnetic dial. In the previous chapter I briefly mentioned the phenomenon of magnetic variation. It is now time to return to this subject in more detail, since it is the reason for the popularity of the magnetic azimuth dial during the latter half of the seventeenth century, and perhaps one of the reasons for its disappearance after the 1690s. It was well known by the sixteenth century that the compass needle did not always point directly to the North, but to a point on one side or the other of true North. The value of the 'magnetic variation' (the difference between magnetic North and true North) was different in different places. Clearly it was necessary to provide compasses with some indication of the value of magnetic North so that they could be correctly oriented. In dial compasses such as those on diptych dials a line was marked on the compass bowl to show the position of the compass needle when the dial was properly aligned. The line would be drawn according to the place of manufacture, but the value of magnetic

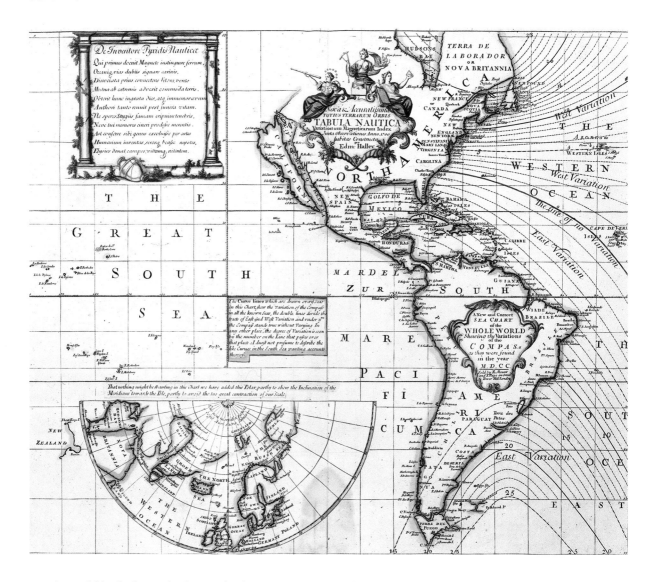

'A General Chart of the Variations of the Compass' (1701), following observations made by Edmond Halley during a voyage to the South Atlantic 1698–1700.

North would be fairly similar for much of Western Europe, and so the inaccuracy of the compass mark would not lead to a great inaccuracy in the time-telling capacity of the dial. Other dials had degree scales marked in the compass bowl so that the user could adjust the position of the compass according to the local value of magnetic North.

Such expedients would have dealt with the problem of magnetic variation, if it were not for the fact that the variation itself varied over time. This was discovered in England during the 1630s, when it was found that measurements of the variation at a point on the Thames had changed by 5° in as little as fifty years and the alteration was not due to inaccurate observations. The phenomenon of the secular variation of magnetic North is now a widely known fact: it is a result of the slow wandering of the Earth's magnetic poles. The accompanying diagram shows how the position of the North magnetic pole has changed through the centuries. Although many dials continued to be produced with the magnetic variation being marked by a line on the compass bowl, it was plain that only those dials which provided

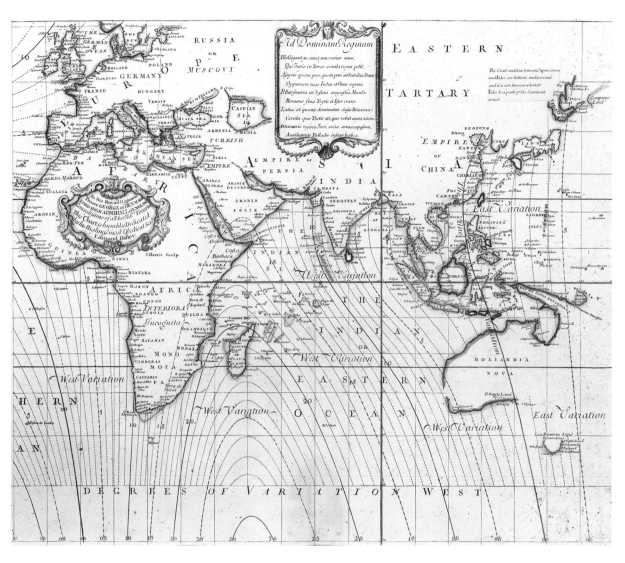

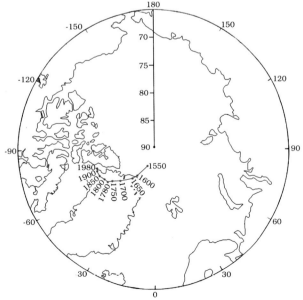

The North magnetic pole has wandered a very large distance in the last 450 years which accounts for the changes in the magnetic variation of the compass during that time.

a degree scale around the compass could be used for a longer period of time and in different places.

During the second half of the seventeenth century, magnetic North in Western Europe was very close to true North. Indeed, the magnetic variation was at zero in London from 1657 to 1661, and in Paris from 1662 to 1665. At this period it would have been possible to make use of a magnetic azimuth dial, which depends on the magnetic variation being zero or very close to zero in order to function accurately. At least, magnetic azimuth dials can be constructed for magnetic variations other than zero, but the procedure is much more complicated, so it is very rare to find one that has been made for anything other than a zero value. Until 1685 (when the magnetic variation had reached 4° W in Paris) these dials would have been acceptably accurate, but after that time they would have lost all usefulness until the next time that the variation approached zero. In fact, the production of these dials did decline during the 1680s, but this may have been as much due to the persecution of Protestants in Dieppe, as to natural phenomena. The Edict of Nantes was revoked in 1685, after which time it was not safe for any Protestant to remain in France.

A Tortoiseshell Magnetic Azimuth Dial

While most of the dials produced in Dieppe were made from ivory, a small number of Charles Bloud's instruments were made from tortoiseshell, as the example here illustrates. This dial comes from the National Maritime Museum collection; similar instruments are held in the British Museum and in the Museum of the History of Science in Oxford. The quality of engraving on this dial is very high and the instrument is not only decorated with flowers such as poppies and lilies, but also with putti, dolphins and mythical beasts. The table of latitudes in the compass bowl consists largely of French towns, as one might expect, but also includes several places in Italy and some of the principal European capitals (Lisbon, Brussels and London).

Dieppe magnetic azimuth dial by Charles Bloud, c.1660, outside of lid.

The main interest in this dial (apart from the unusual material) lies in the engraving on the outer face of the upper leaf. Here we find the letters 'IHS' engraved between a cross surrounded by a sunburst and a heart pierced by three arrows; the engravings are set within a circle of flames. All of these motifs are symbols of the Jesuit order, the initials 'IHS' being one of the mottoes of the Jesuits. The standard of the engraving shows that this dial must have been made for a high-ranking member of the Jesuits.

The Society of Jesus was founded by Ignatius Loyola in 1534, being granted status as a Roman Catholic religious order in 1540. From the start there was a strong emphasis on both education and teaching: the Jesuits were responsible for five hundred colleges by the early seventeenth century and their missionaries went to North and South America, India and the Far East. Jesuits working in China and Japan brought knowledge of Western scientific and technological practice to these areas, including the use of portable sundials. This side of their learning was particularly welcomed by

Outside of base, showing calendar volvelle.

Inside, showing the lunar volvelle and the main sundial.

the Ming dynasty in China, which granted the Jesuits the protection of the Imperial Court, and permitted them to continue their teaching and missionary work for many years. The Jesuit order was suppressed in the later part of the eighteenth century, but revived in 1814 and continues to be an important movement in the Roman Catholic Church.

It seems strange that a Huguenot dial maker, who was later to be persecuted by the Catholic government of France for his faith, should be making dials for a Catholic Order. However, this is simply one more example of the importance of patronage to the trade in sundials and other scientific instruments. Such detailed and carefully engraved sundials would have been costly to buy and would have brought in much necessary income for the maker. Any maker who wished to be able to rely on a continued market for his wares could not afford to be particular about those from whom he received patronage, nor could the purchaser be particular about the maker if he wished for the best quality dials. An important patron could also bring custom to a maker by recommending the craftsman to his friends; he was to be valued and not to be rejected on the basis of religious, political or any other differences. So we see that while the Protestant-Catholic antagonism was strong it did not prevent trade and other links between the two denominations, unless political affairs intervened, as they did during the 1680s after the revocation of the Edict of Nantes. Huguenots were persecuted once again, and those who were not killed fled to the safety of Protestant countries such as the Netherlands and England, leaving Dieppe as a Catholic town once more.

Augsburg

View of Augsburg by F. Garden in Salmon's Universal Traveller, 1752

Moving inland some five hundred and fifty miles from Dieppe, we now return our attention to Augsburg. Since the time of the Schisslers the city had undergone some changes and had been hit particularly hard by the Thirty Years War. This war was the result of both religious conflicts among the states of the Hapsburg Empire and the desire for a greater degree of autonomy for the German lands. At the outset of the war in 1618 the main protagonists were from the central European lands, but it gradually engulfed most of the countries of Western Europe. Augsburg was occupied not only by Imperial and Bavarian troops, but also by the army of the King of Sweden, Gustavus Adolphus, who fought for the rights of the Protestant states of Germany while ensuring the safety of his own country. During the early part of the war the Imperial armies had the upper hand and the Edict of Restitution (1629) effectively banned Protestant worship in all areas of the Empire. However, the intervention of Sweden finally brought a new settlement in 1648 (the Peace of Westphalia) which split the Empire into many individual states, each of which had the right to choose its state religion; Protestants were finally set on an equal footing with Catholics and religious war ceased to be a problem within the Holy Roman Empire.

Following the close of the Thirty Years War, Augsburg (which remained an imperial city) was able to regain stability and over the next hundred years it became a noted centre for art, especially that springing from the new Rococo movement. In particular, the excellence of the work of its gold- and coppersmiths became known throughout Europe. It is scarcely surprising, therefore, to find that the same hundred years were the heyday of the sundial makers of Augsburg.

The Augsburg Dial

The distinctive Augsburg dial developed from earlier equinoctial dials, and was similar to those which had been incorporated in the astronomical compendia of the sixteenth and early seventeenth centuries (as, for instance, the one on the Cole compendium discussed in chapter four). The dial almost always consisted of an octagonal base with a large compass set in the centre of the base plate. The equinoctial ring of the dial was hinged to the base plate and its gnomon was a narrow rod set at the centre of a pivoted bar lying across the East-West diameter of the hour ring. Hours were usually marked from 3am to 9pm around the equinoctial circle, but occasionally the full twenty-four hours were inscribed. The latitude of the dial was set by means of a latitude arm hinged to the base plate, against whose scale the angle of the equinoctial ring could be set; the ring was then held in place by the friction between the two pieces of metal.

These were the main constituents of the Augsburg dial, but, as one might expect, there were often variations. Some instruments have square base plates instead of octagonal ones; some include plumb bobs to ensure that the dial is set upright (it can often be adjusted by means of levelling screws at the corners of the base plate). Some carry latitude tables on the underside of the compass bowl, while others incorporate the table into the lid of the box; still others came with paper tables of latitudes, many of which have now been lost.

The octagonal form of equinoctial dial was also used occasionally by French makers. Here we see a typical Parisian 'Augsburg' dial by Michael Butterfield. Like other French versions, it is heavier than its German counterparts. It also carries the characteristic gnomon. The hour ring and gnomon of this form of Augsburg sundial now appear as the emblem of the British Sundial Society: the emblem was chosen because it is based 'on the one class of dial which is fundamental to the understanding of the whole art of dialling, the *equinoctial* sundial'.

French form of Augsburg dial by Michael Butterfield, c.1700.

The Augsburg Makers

Production of the Augsburg dial differed from that of the magnetic azimuth dial in that versions of it were made for both ends of the market. So while there are many highly crafted instruments, other makers created a mass market in cheap dials. These cheap dials were made of brass and were often rather poorly executed, with crude designs and a lower accuracy than their finer cousins. They were the products of a group of makers who appear to have specialised in catering for this particular end of the market. Chief among the makers were the brothers Johann Georg and Andreas Vogler, who were sons of a miller from a nearby district. Johann Georg settled in Augsburg during the 1740s, establishing a workshop where he worked until his death in 1765. The following year, his brother Andreas took over the shop and the trade, continuing in business until the 1790s; he died in 1800. The dynasty continued through the marriage of Andreas' daughter, Maria Katharina, to Johann Nepomuk Schretteger in 1797. Schretteger was also a dial maker, and he carried on the manufacture of Augsburg dials until his death in 1843. The other two main makers of cheap dials were Ludwig Theodor Müller (c.1710–70) and Lorenz Graßl (c.1740–1805). While the community of makers was not quite so close-knit as that of the Nuremberg diptych-dial makers, there was clearly still a certain tendency in Germany to keep the trade 'within the family'.

All these cheap dials had sprung from imitation of the more highly crafted Augsburg dials which were designed at the end of the seventeenth century, and through which the Augsburg form first came into being. The principal makers of these dials were Johann Martin (1642–1721) and his relation Johann Martin Willebrand (1714–42), and Nikolaus Rugendas. Johann Martin was born in Frankfurt in 1642 and had moved to Augsburg by 1669 when he married Maria Barbara Weckerlin. He specialised in portable sundials, which all display a high standard of workmanship. In particular, he was responsible for the design of the crescent form of the equinoctial sundial, a relative of both the Augsburg dial and also the universal equinoctial ring dial (like the latter it is self-orienting, but the form more closely resembles the Augsburg dial). Martin was joined by his step-brother, Johann Matthias Willebrand around 1682. Both Willebrand and his son, Johann Martin Willebrand, were skilled dial makers. I have not given any dates for Nikolaus Rugendas, because there were three makers in Augsburg with that name and it is not easy to tell which was responsible for the sundials signed by Rugendas. It was probably Nikolaus III, since the others were mainly active as clockmakers. Nikolaus III was born in 1665 and died in 1745, and it is probable that he had some links with Johann Martin, since his dials are of a very similar style.

A Dial by Johann Martin

This example of Martin's work demonstrates clearly the superiority of his skill as an engraver. The simplicity of the dial's design is set off by the fine

Augsburg dial by Johann Martin, late seventeenth century.

Back of the dial with projection of the celestial sphere.

Wind ring.

Wind vane for use with the wind ring.

detail of the decoration. In addition to the normal parts of the dial, the front of the base plate carries a date scale and a zodiac scale which can be used for finding the sun's place in the zodiac for any day of the year. The zodiac scale is particularly finely engraved, with each sign being shown by a delicate drawing rather than the more usual symbol. On the reverse of the dial plate we see three small volvelles, for determining the age of the moon, the number of hours since sunrise or sunset, and the day of the week, given the date. In the centre of the base is a projection of the celestial sphere, which can be used for finding the declination of the sun (its distance from the celestial equator) at any time of the year, and the hours of sunrise and sunset, among various other functions.

This dial also comes with a number of accessories. The accompanying silver disc carries a nocturnal and lunar volvelle on one side. Unlike the nocturnals on Gunter quadrants, this one does not show the constellations; instead it marks the positions of particular stars, as lines from the centre of the disc to the hour at which they are in the south of the sky. Otherwise its use is much the same as before. The back of the nocturnal has a central wind ring which can be used with the beautifully crafted wind vane to find the direction of the wind. This disc also carries a set of hours, and it can be rotated over a disc which is marked with degrees and various places according to their longitudes. Zero longitude is given as the western edge of the Canary Islands, a common choice at that period. The disc allows the time

Nocturnal from the Martin dial.

of day to be found in any of these places around the world, if the time is known in any one of them (a feature which would be useful today, but which certainly cannot have been in the late seventeenth century!). Finally, the case itself supplies an extensive latitude table on the disc set into the inside of the lid.

In stark contrast to the Martin dial, the second example – an instrument by Andreas Vogler – has no elaborate decoration, is small and not particularly accurate. This is typical of the instruments which were made for the mass market. The magnetic variation is indicated by a single line on the

Detail of case, showing table of latitudes.

Augsburg dial by Andreas Vogler, second half of eighteenth century.

compass (preventing the instrument from being used at a long distance from Augsburg or many years after its production). The decoration is rather crude, as is the positioning of the hours on the equinoctial ring. Both East and West points of the compass have been given as 'OR' (for *Oriens* – East); the West point should have been identified as 'OC' (*Occidens*). On the back of the dial plate, a short table of latitudes is provided giving details for Augsburg, Paris, Krakow, Prague, Leipzig, Cologne and London. This may have been supplemented by an accompanying printed table of latitudes, but this has since been lost.

The Butterfield Dial

The third type of dial to which we are turning our attention in this chapter is less restricted to a particular locality, although it was most commonly made in Paris, between about 1675 and the end of the eighteenth century. It owes its name to an Englishman who established a workshop in the French capital in the latter part of the seventeenth century. Michael Butterfield was born in England in 1635, but moved to France around 1663. The first certain mention of him is in 1677 when he had a workshop '*Aux Armes d'Angleterre*' (at the sign of the arms of England) in the Faubourg

Circumferentor by Michael Butterfield, c.1690.

Saint-Germain, Paris, and was producing scientific instruments. His skill and ingenuity soon brought him into contact with the Académie des Sciences (the French equivalent of the Royal Society) and he was commissioned to produce work by various of its members.

Paris at the end of the seventeenth century was rapidly rising to become the greatest city in Europe, eclipsing the Imperial cities of Vienna and Prague. Under Louis XIV (the 'Sun King') and his chief ministers, Cardinal Jules Mazarin and Jean-Baptiste Colbert, France's capital overcame the years of poverty caused by the religious wars of the sixteenth and early seventeenth centuries, and the devastation of the Fronde, the civil war of 1648–53. Money was lavished on extending the paved streets in the city, providing streetlighting and night watches to patrol the streets, and improving the water system. The Académie des Sciences was founded in 1666, to become a centre for scientific discussion and investigation; this was soon followed by the construction of the Paris Observatory, the first permanent site for the study of astronomy in France. Both these creations were of vital importance to the instrument-making community in Paris, since many

Horizontal dial by Pierre Sevin,
late seventeenth century.

more instruments were now required by the members of the Academy and the astronomers at the Observatory. High-quality precision instruments were especially in demand and those makers who could supply them prospered. Among these men was Michael Butterfield.

Butterfield's workshop became one of the most substantial in Paris, and his output was extensive, including surveying instruments, drawing and calculational instruments and, of course, sundials. In 1717 he was honoured by a visit from Peter the Great of Russia, who was then touring Western Europe. Nearly twenty years before that he had, very remarkably, been even more greatly honoured by the grant of a coat of arms, a privilege unique among the Parisian makers, as far as is known. He died on 28 May 1724.

The essentials of the Butterfield sundial are a dial plate engraved with a number of different hour scales serving different latitudes, a compass, and an adjustable gnomon. These are all present in the dials which are signed by Michael Butterfield himself. However, it was not Butterfield who originally designed these instruments. Examples are known which clearly date from before the time that Butterfield began working in Paris and they were probably developed from a contemporary form of horizontal dial (illustrated above) since many of the features were the same. The earliest known dated Butterfield dial is one made by Roch Blondeau and dated 1673 (see page 104), four years before Butterfield was known to be working in Paris. It is likely that Timothé Collet was also producing this type of dial before Butterfield started to work in Paris. Why the dial became known as a

Butterfield dial by Roch Blondeau, 1673.

'Butterfield' dial is unclear, but it may be due to Butterfield's regular inclusion of a bird-shaped pointer on the gnomon, whose beak was used to mark the latitude for which the dial was set. This bird became a standard part of all Butterfield dials produced in the eighteenth century.

Interestingly, the form of the bird changed very little throughout the century and a quarter during which the Butterfield dial was produced. Other ornamentation on the dial would vary from instrument to instrument and from maker to maker, and differences would inevitably be found in the choice of latitudes for the hour scales and in the list of towns in the table of latitudes which was almost always inscribed on the reverse. However, the bird index itself hardly varies and seems to have been seen as the distinguishing characteristic, not simply in form but also in content of decoration and shape.

The Butterfield dial is a type of semi-universal horizontal dial, that is, a horizontal dial which can be used in a number of different latitudes, but not in the full range from 0° to 90°. It is generally either octagonal or oval in shape and is almost always marked with three or four different hour scales, each of which is marked with a specific latitude: as we have already found, the hour angles change for different latitudes, and therefore one scale is not sufficient for every type of dial. Usually these hour scales are set for latitudes of 3° to 5° apart, and for latitudes lying between those specified the nearest hour scale is used. In order for the dial to be used in different latitudes the angle of the gnomon must be adjustable. The gnomon proper is marked with a latitude scale and it can move within the bird index which marks the latitude angle against the scale.

Despite the beauty of the Butterfield dial it was not a particularly practical instrument. In the later years of the eighteenth century it was criticised by one of the leading writers on dials. Dom. François Bedos de Celles in his *La Gnomonique Pratique* (*Practical Dialling*) declared that the compass was too small and could not be altered for magnetic variation (although this is

Butterfield dial by Timothé Collet, mid-seventeenth century.

Butterfield dial by Michael Butterfield, c.1700.

slightly unfair – some of the compasses were marked with degree scales so that the variation could be taken into account), and that the dial plate was difficult to read because of the large number of scales on a small instrument. Yet they were extremely popular, judging by the number which survive. There are several possible reasons: it may have been difficult to buy other forms of portable dial in Paris; they may have been well marketed by shrewd makers; or it may simply be that they were fashionable and so people bought them.

Two English Butterfield Dials

I have chosen two atypical instruments to illustrate this type of dial. They are atypical in that they were made by an Englishman – John Rowley – and also in their style of decoration and the inscriptions on the reverse. Both were made as presentation pieces for wealthy patrons, either to be given by the instrument maker himself, or by a friend of the recipient. Each one is personalised to fit the original owner, but otherwise they are the same in shape, size and the layout of the hour scales and compass.

The first dial was made for the fourth Earl of Orrery, as we can tell from the inclusion of his arms on the front of the dial plate. The Earl was an important patron of natural philosophy at the beginning of the eighteenth century and had a large collection of scientific instruments. He is best known for having given his name to the orrery, a form of planetarium which demonstrated the movements of the planets around the sun; John Rowley was responsible for the manufacture of many of the grandest of these instruments. The fact that this sundial was not part of the Earl's collection of instruments (now housed in the Museum of the History of Science in Oxford) suggests that it was regularly used for timekeeping, and was not kept as a scientific curiosity.

The dial carries many typical features of a Butterfield dial, including the bird gnomon; however, the table of latitudes on the reverse has been replaced by a table for the equation of time, with values given for the tenth, twentieth and thirtieth of each month. This was an important set of data necessary for calibrating clocks and watches from sundials. We have seen that both the height and the direction of the sun at a particular time of day vary through the year. We now have one last factor to add to the movement of the sun. It is only true to say that the sun is always due South at midday if days are measured by the sun. If an accurate clock or watch is used to measure the day it will not always show twelve o'clock when the sun crosses the meridian line. Sometimes it will appear faster than the sun and sometimes slower. This alteration in the length of each day is the result of the Earth's unequal distances from the sun through the year. It was not important when time was measured by sundials, but the advent of clocks (which mark mean (i.e. average) time, rather than solar time) necessitated an accurate measurement of the difference between solar time and mean time on each day of the year. Accurate tables were first compiled in the late seventeenth century and appeared on some sundials after that.

The second dial was made for the first Duke of Marlborough. Like the previous one it carries its owner's coat of arms on the front of the dial, but the personalisation is here extended by the use of a griffin – the supporter for the Duke's shield – as the index for the gnomon, with the forked tongue of the animal marking the latitude. It is one of the few examples of a Butterfield dial where the bird has been replaced by another indicator. The reverse of the dial carries a latitude table which includes a very different list of towns from the standard one: The Hague, Utrecht, Berlin, Antwerp, Maastricht, Cologne, Dunkirk, Venlo, Frankfurt, Mainz, Liège, Luxembourg, Triers, Metz, Thionville, London and Paris. These places were chosen to commemorate Marlborough's successful campaigns in Germany and the Low Countries, particularly during the War of the Spanish Succession.

At the end of the seventeenth century there was a crisis in Europe because the King of Spain, Charles II was dying and had no child to succeed him. There were three claimants for the Spanish throne – Louis XIV's son, whose mother was Charles' sister; the Holy Roman Emperor, Leopold I, who was married to another of Charles' sisters; and Joseph Ferdinand, the Elector of Bavaria. When Joseph died in 1699, the problem of finding an acceptable successor became acute, since the tensions between the remaining claimants were high. When Charles made a will in favour of Philip of

Charles Boyle, 4th Earl of Orrery, artist unknown; copy after original by Charles Jervas (1707).

Detail of the gnomon from the Marlborough dial, showing the griffin index.

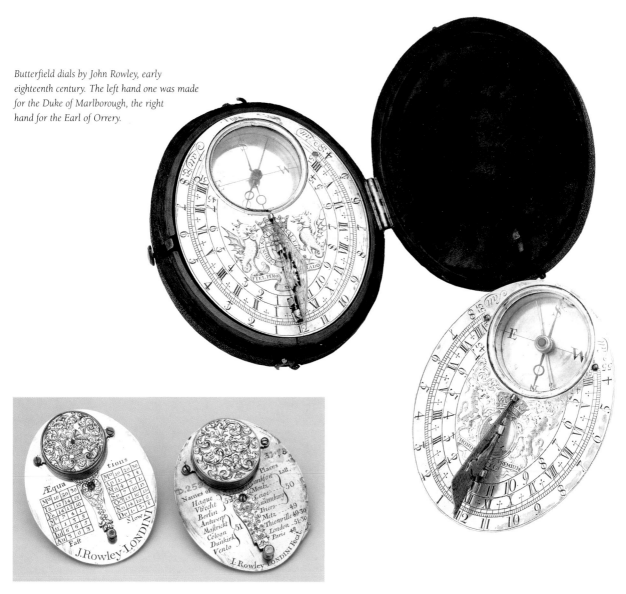

Butterfield dials by John Rowley, early eighteenth century. The left hand one was made for the Duke of Marlborough, the right hand for the Earl of Orrery.

Undersides of the two dials, with the Orrery dial on the left.

Anjou, Louis XIV's grandson, it was clear that war was inevitable. The might of France was pitted against a Grand Alliance consisting of England, Holland, the Holy Roman Empire and several of the German and Italian principalities. John Churchill, the Duke of Marlborough, was Commander-in-Chief of the British Armies when war was declared in May 1702.

The early years of the war were occupied with capturing towns by siege. Liège and Venlo both fell to the Allies in 1702; Triers (or Trèves), which had been seized by the French in 1701, was recaptured in October 1704. The year 1704 saw Marlborough's great march to the Danube, which ended in the Battle of Blenheim; Cologne, Frankfurt and Mainz were important supply posts for this march, while Maastricht was one of the centres of Allied activity.

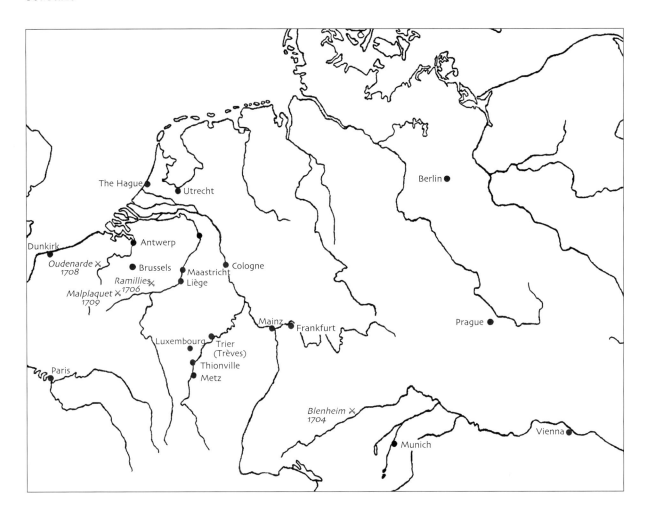

The campaigns of John Churchill, 1st Duke of Marlborough.

(Marlborough had also been present at the siege of Maastricht in 1674, during the Franco-Dutch War of 1672–78 and had shown his military genius there at a young age.) After the Allied victory at Ramillies in the spring of 1706, Antwerp capitulated rapidly. The remaining places listed on the dial did not change hands, but are significant in other ways. Dunkirk, Luxembourg, Metz and Thionville were all French outposts; The Hague and Berlin were the capital cities of Britain's allies, Holland and Prussia; the treaty which finally ended the war in 1713 was signed in Utrecht. So this dial commemorates the scenes of the Duke of Marlborough's victories, and was probably made as a triumphal memento after his return to England.

The eighteenth century was probably the culmination of the art of the sundial maker and the heyday of sundial use. We have seen that many fine examples of the craft were being produced from 1650 onwards, and the numbers which survive from this period indicate a greatly enlarged market for dials. However, by the beginning of the nineteenth century the watch was beginning to supplant the sundial as a timekeeping device and in the next chapter we will see how the sundial was affected by this.

Chapter Seven

WATCHING THE TIME

Introduction

The mid-eighteenth century was the peak of the sundial trade. However, there were still some important innovations before sundials were completely replaced by watches as the most common everyday timekeepers. These innovations are not particularly important for increased accuracy but simply for new designs of dial. Indeed, in terms of accuracy, there were few portable dials produced in the late eighteenth and early nineteenth centuries which could improve on the universal equinoctial ring dial, one of the most accurate portable dials produced before the beginning of the twentieth century. We will begin by looking at the new designs of this period, before finally turning our attention to the rise of the watch and its supplanting of the dial as the favoured timekeeper. However, we will see that, even in the mid to late nineteenth century, the dial had its part to play as a regulator for the watch.

Table Dials

In the second half of the eighteenth century the equinoctial dial, previously most popular in the form produced in Augsburg, began to be manufactured regularly by English makers, in a rather different form from any that had previously been developed. In addition, the same makers began to produce portable inclining dials. The latter look very similar to the equinoctial dials but work on rather different principles, so we will return to them after first considering the equinoctial dials.

Both forms of dial have a circular base plate which contains a compass. The compass almost invariably carries two bubble levels set at right angles to one another beneath the compass needle. These levels are used to ensure that the dial is set flat on a surface – three levelling feet at equal distances apart around the base plate can be adjusted until the bubble levels indicate that the compass is horizontal. The levels provide a rather more accurate form of levelling indicator than the plumb bob which had been the standard means before this time and which can be seen on many earlier

instruments. The compass carries a degree scale around its perimeter, which allows compensation for the magnetic variation to be made accurately.

Hinged to the North side of the base plate is the hour plate, just as on an Augsburg or other early equinoctial dial, and a latitude arm is set on the East side of the plate. Once again, these features are the same for both equinoctial and inclining dials, but here the similarities end. As one would expect, the equinoctial dial carries an hour scale which is divided evenly around the circle from three or four o'clock in the morning to eight or nine in the evening. As with the Augsburg dials the gnomon is provided by a rod (here somewhat thicker) set on a pivoted bar. Because the gnomon is of an appreciable thickness the hour scale is divided into two parts and it is only the edge of the shadow of the gnomon which is used to mark the time (one edge for morning hours and one for afternoon hours).

The advantage of these instruments over earlier forms of equinoctial dial is their accuracy. They are typically divided at five-minute intervals and the accuracy of the divisions may well have been increased by the use of a dividing engine, rather than the eye of the craftsman alone. The dividing engine was developed by the English maker Jesse Ramsden, in the mid-eighteenth century, and it allowed any circle to be divided into degrees and parts of a degree with a far greater accuracy than could be achieved by eye alone. It was commonly used for dividing the scales on sextants and astronomical measuring instruments, but it seems quite likely that it could have been employed for dividing the degree and hour scales on equinoctial dials.

Equinoctial dials by Andrew Pritchard (left) and Dring and Fage (right), first half of the nineteenth century.

The inclining dial is similar in appearance, but has some important differences. The main one is that it is not an equinoctial dial but a horizontal dial. This might seem a bit surprising at first for an instrument that is designed to be used in a large number of latitudes – a horizontal dial with an angled gnomon is surely constructed to be used in one latitude only. This is true, but if that dial is taken to another latitude it can be tilted so that its gnomon still follows the axis of the Earth and its hour plate is parallel to the horizon plane of the place for which it was made. Set up like this, it will continue to tell the time correctly, even though it is not in its original location. An inclining dial follows this principle and can be adjusted for a large range of latitudes. They are usually designed for a high latitude and then can be tipped up so that they are suitable for all latitudes below that for which the plate is engraved. The picture overleaf shows an inclining dial carrying a standard gnomon for a horizontal dial and a dial plate divided according to a high latitude (60° N). As the hour plate is raised the angle of the gnomon lowers (the latitude is decreased) and the dial can be used for telling the time at points further South than 60°.

Although a few instruments working on this principle were produced

Jesse Ramsden, Optician to his Majesty (1735-1800). Mezzotint produced by Robert Home (artist), John Jones (engraver) and Molteno,Colnaghi & Co (publishers) 1 January 1791.

Inclining dial by Cox, late eighteenth century.

during the early years of the eighteenth century, inclining dials only became common towards the end of that century, when English instrument makers produced a standard design which continued to be made during the early years of the nineteenth century. The kinship between these inclining dials and the equinoctial dials being produced at the same time suggests that one was developed from the other (most likely the inclining dial from the equinoctial dial, but we cannot be sure).

It cannot really be said that either of these forms of sundial could be carried in the pocket, as was the case with many previous dial types. However, they are clearly portable, and the inclusion of universal latitude settings indicates that they were made for use by people who would be travelling. The high quality of the materials and the standard of the engraving suggest that these were expensive instruments only made for the well-to-do. They could be set on a table in a sunny window and so tell the time, being moved from room to room as the sun moved. It is possible (though it is difficult to find evidence for this) that they were taken on trips abroad and used to provide a relatively accurate timekeeper when the family clock was not available.

Inclining dial inscribed in Arabic, late eighteenth century.

An Islamic Inclining Dial

Butterfield dial made for the Arabic market by Nicolas Bion, c.1700.

While most of these equinoctial and inclining dials were made for a European market, some examples survive of instruments which were sold in Arabic countries. There are several known inclining dials inscribed with Islamic characters, of which this is one example. It may well have been made in England, where the divisions for the time scale and the degree scale in the compass would have been engraved. However, it would then have been transported to Turkey or perhaps Persia, where the inscriptions would have been added. The characters used to mark the hours are unusual forms of Arabic numerals: they are found on Islamic clocks and other sundials in imitation of the Roman numerals which were popular on both clocks and dials made in Europe. The more normal form of Islamic numbering can be seen on the scale marking the minutes and on the latitude arc.

Most of the remaining inscriptions on the dial (on the base, the edge of the base plate, the underside of the hour plate, etc.) are a table of latitudes which gives not only the latitude of each town or city but also the *inhiraf* and *jifa*. These are used for finding the direction of Mecca from the town where the dial is in use. This is of particular importance, since a Muslim must face Mecca when praying; so the dial not only indicates the time

Underside of the inclining dial showing part of the table of latitudes.

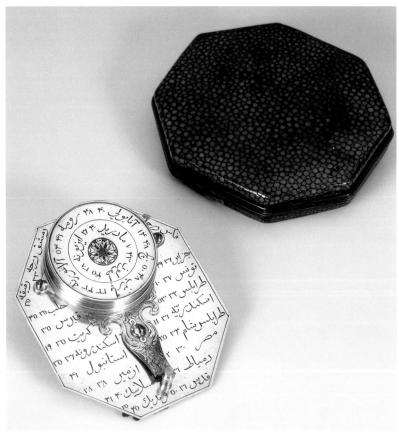

Table of latitudes on the back of the Bion dial.

114

Magnetic dial by Richard Ebsworth, mid-nineteenth century.

(providing a reminder of the five times of prayer), but also marks the direction of Mecca for use during those times of prayer.

It was a relatively common practice for dials made in the West to be used in Arabic countries. This Butterfield dial was made by Nicolas Bion and is in many respects the same as those which he made for sale in France. However, all the markings are in Arabic and the towns which are listed on the reverse are places in Turkey, Persia (now Iran) and so on. Just as the European Butterfield dials alternate Roman and 'Arabic' numerals for the different hour scales, so this instrument alternates the standard Islamic characters and the more formal ones developed as an imitation of Roman numerals.

Magnetic Dials

While the equinoctial and inclining dials were being developed for the upper end of the market, a rather different form of dial was invented for the lower classes. This type of dial was certainly a poorer timekeeper than many of the contemporary dials and probably most watches. However, it was probably cheap and certainly very easy to use.

This type of dial is known as a magnetic dial and it consists simply of a

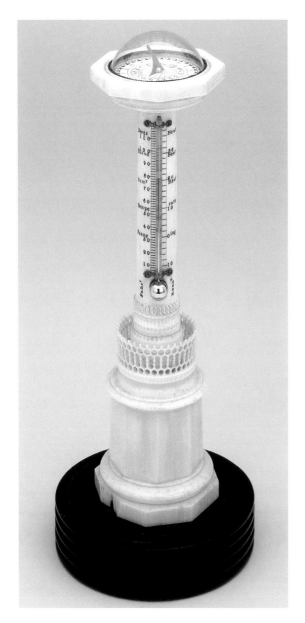

Magnetic dial with thermometer, English, mid-nineteenth century.

Magnetic dial with thermometer, by Thomas Staight, mid-nineteenth century.

dial plate, to the underside of which is fixed a magnetic needle. The plate is suspended on a point and it will automatically turn itself to the correct position. So, unlike other dials where the user has to rotate the instrument until the needle is in the correct position relative to the meridian line of the dial, this dial is in position as soon as the needle has stopped moving, as long as it is set on a horizontal surface. Such ease of use would go some way to balancing the fact that the magnetic dial was not at all accurate.

The reasons for the lack of accuracy were several. First, the instruments were small, so that they would not overweight the needle and hinder its motion. Secondly, they were subject to the problems caused by magnetic variation. Some of these could be solved by setting the needle at the correct angle to the meridian line when it was first mounted on the card. However, the fact that magnetic variation changes from place to place and also over time meant that the dial was bound both by location and by period. It could only be used within a small radius of the place for which it was made and moreover could only be used for a period of about ten years, before it became too inaccurate to provide a reasonable indication of the time. It seems ironic that some of these magnetic dials include tables giving the equation of time – such inaccurate instruments could never have been used for checking clocks and watches for accuracy.

Many of these dials were made as simple instruments for cheap sale. However, they were also incorporated into more fanciful creations. Objects such as this pillar which carries a dial on the top and a thermometer on the side were made of costly materials (in this case ivory) and can only have been intended as novelty items for the rich. They may well never have been used for their supposed purpose (though their temperature indicators were likely to be more accurate than those for time), simply being acquired for their ornamental value. Sundials are also found combined with barometers and with rulers (though the latter instrument does not normally carry a magnetic dial).

A good example of such usage of the magnetic dial is this combination instrument made by Thomas Staight of London in the second quarter of the nineteenth century. The central dial is set in an ivory plate with a thermometer wrapped around it. The thermometer gives scales in Fahrenheit and also in Réaumur. The latter temperature scale is calibrated so that the freezing point of water is 0° and the boiling point is 80°. Various points on the scales are named with appropriate comments: 'Freezing' at 32°F and 0°R, 'Temperate' at 55°F (10°R), 'Summer heat' at 76°F (19.5°R) and 'Blood heat' at 98°F (29.5°R). The sundial itself is small and would not have been very accurate, even when first made.

Watches

During the nineteenth century the portable sundial began to fall out of favour, as it lost place to the watch as the preferred method of personal timekeeping. While hundreds of sundials survive from the seventeenth and eighteenth centuries, the numbers are considerably lower for the nineteenth century, while at the same time the watch became a commonplace item.

However, watches had been available since about 1500. Why was it that it was only in the nineteenth century that they posed any threat to the supremacy of the portable sundial? The answer lies in the relative accuracy of the two methods of timekeeping. Early watches were inaccurate and were also expensive. Hilaire Belloc's epigram was an unfair and indeed unrepresentative description of the early relationship between sundials and watches:

Gold pocket watch by Benjamin Lewis Vulliamy, c. 1847.

I am a sundial and I make a botch
Of what is done much better by a watch.

A better understanding of the relationship can be found in the works of William Emerson, in his *Dialling. Or the Art of drawing Dials, on All Sorts of Planes whatsoever* (1770). He writes that 'And tho' we be furnished with some sorts of moving machines, which will do this [tell the time], as clocks and watches, yet these are often out of order, apt to stop and go wrong, and therefore require frequently to be regulated by some unerring instrument, as a dial; which being rightly constructed, will always (when the sun shines) tell us truth. And therefore whether we have any clocks or not we should never be without a dial.' The belief in the superiority of the sundial over mechanical timepieces is evident.

It was only as means were found for improving the accuracy of watches and as they became cheaper to make and to buy that they began to be sold in numbers approaching those of sundials. In order to understand this change we need to look briefly at the development of the watch from its original invention through to the changes made in the late eighteenth and early nineteenth centuries which so greatly increased its accuracy and made it the preferred timekeeper rather than the portable sundial. Many books have been written on this subject, and I only intend to give a short overview here.

Watches were first produced about two centuries after the first clocks. Those early clocks had mechanisms which were not suitable for use in watches because they required weights on long ropes to provide the driving power for the movement of the clock. A suitable drive mechanism for the watch was introduced with the invention of the mainspring, where the energy held in a coiled spring is used to drive the mechanism. These springs were first used in clocks in about 1430 and it seems that watches were developed fifty to seventy years after that time.

The main parts of a watch are the driving force (provided by a mainspring), the gear train which moves the hand or hands on the dial of the watch and the mechanism which links driving force and gear train together. This last is known as the escapement and it regulates the transmission of force to the gear train. The illustration here shows the early form of watch mechanism, with what is known as a verge escapement. There were various problems with this type of mechanism, some arising from the escapement, some from the mainspring itself. Taking the latter first, we find that there are problems with using a spring to provide a constant power, because it has a very high power output initially which then decreases quickly to a level that descends gradually and evenly for a few hours, after which point it then decreases very rapidly to reach a stop. This is not much use for constant timekeeping and so a device known as a fusee was added to compensate for the alterations in the power provided by the mainspring. The fusee and mainspring together formed the standard motor for watches until the nineteenth century and were still being used in England as late as the beginning of the twentieth century.

Problems with the escapement are basically twofold. Firstly the friction

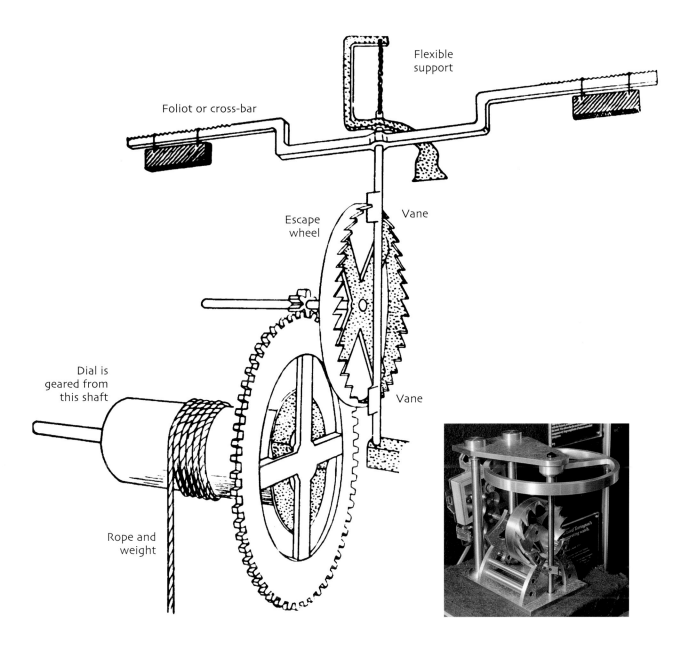

Flexible
support

Foliot or cross-bar

Escape
wheel

Vane

Vane

Dial is
geared from
this shaft

Rope and
weight

A verge escapement in operation.

between the verge pallets and the escape wheel interferes with the accuracy of the watch. Secondly, the recoil created when the pallets of the verge shaft hit the escape wheel is not always constant and so means that the lengths of time between one pallet and the next touching the escape wheel cannot be kept constant.

Two final problems with these early watches were the poor quality of the lubricants for keeping the parts greased and the uneven nature of early springs. All these problems meant that watches kept very poor time and were normally only fitted with hour hands and marked out to quarter hours, because they could not be expected to be more accurate than that in telling the time.

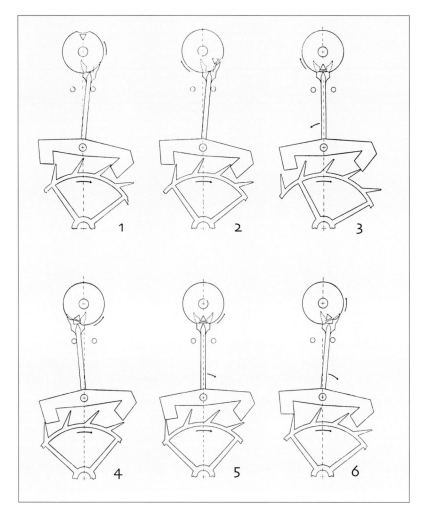

A lever escapement in operation.

The first major improvement in the timekeeping abilities of the watch came with the invention of the balance spring in 1675 (separately claimed by Robert Hooke and Christiaan Huygens). The purpose of this spring was to try to hold the balance wheel in its rest position and so to provide a force which would always bring the balance wheel back to the centre of its swing (just as gravity acts on the pendulum of a clock). The balance spring helped to regulate the movement of the balance wheel and was so effective that it improved the timekeeping of watches enormously: some were accurate to within two minutes per day. However, the cost was still high and ordinary people would still have been more able to afford a sundial than a watch. The introduction of the balance spring also made it apparent for the first time that temperature had an effect on the elasticity of the mainspring and so affected the accuracy of the watch.

During the eighteenth century numerous attempts were made to improve the accuracy of watches, particularly in order to produce an accurate marine chronometer, but for domestic use as well. The invention of the bimetallic strip overcame the problems caused by temperature change. There were also improvements in the quality of the wheel work in the gear

Compass dial by Rousseau, second half of the 18th Century. In an interesting side development, sundials were made which resembled the fob watches of the time. This dial mimics the shape of a watch, while being a fairly simple form of dial. The reverse carries the portrait of a lady, perhaps the Comtesse d'Angevilliers, whose husband has been suggested as the sitter for a very similar dial, sold at Christie's in 1988. These watch-type sundials were mostly produced in France, in the second half of the eighteenth century.

Dipleidoscopes by Edward Dent, c.1850.

chain and frequent cleaning and replacement of the lubricant removed many of the problems created by dirt. However, it was work on different forms of escapement which really revolutionised watch production.

Numerous different types of escapement were designed, but the one which was to become the standard form was the lever escapement, invented by Thomas Mudge in 1770 (though it is possible that Julien le Roy also had a claim on the design). As the diagram on page 120 shows, this new form

Dipleidoscope by Edward Dent, 1850s. Unlike most dipleidoscopes, this one can be used across a range of latitudes.

of escapement separated the balance and balance spring from the rest of the watch movement, so allowing these parts to oscillate freely and not to be affected by friction between the parts, except at the moment when the impulse is transferred to the escape wheel. As the balance swings the lever arm is unlocked and one of the lever pallets comes into contact with the escape wheel giving it an impulse. The balance wheel then swings the lever back to its central position and on until the second pallet engages the escape wheel, moving the latter round once again. It is the lack of interference with the movement of the balance which makes the timekeeping reliable and increases the accuracy of the watch.

Early forms of the lever escapement were delicate and so could not be subjected to rough handling in a watch. They were also not originally popular with watchmakers and were only widely developed from the second decade of the nineteenth century. However, from that point on the lever escapement became widely used as the standard form of escapement in a watch and it is the escapement which will be found in all mechanical watches produced today. At the same time improvements were made to the shape of the balance spring and the fusee was commonly replaced by the going barrel – a more accurate means of regulating the mainspring.

All these developments led to a much greater increase in the accuracy of the watch. At the same time, mass production of watches began to become a reality. Both Switzerland and America developed factory production of watches which reduced prices and hugely enlarged the numbers of watches available for purchase. English watchmakers continued to work in small companies but even here the introduction of batch assembly increased the output of the workshops considerably.

Demographic changes brought about by the industrial revolution further contributed to the changing methods of timekeeping. As people moved from a rural to an industrial base the working day was no longer ruled by the hours of light, and the factory system generally required everyone to work at the same time. Meanwhile railways were developing and these really did necessitate owning a portable timekeeper which could be relied upon to give the time to the nearest minute (since railway timetables were given to this accuracy). With such demands for accuracy in timekeeping it was inevitable that the watch would become the preferred method of finding the time and that the portable sundial would fade from view.

Sundials as the Watch's Assistants

Even when the watch took over from the sundial as the preferred portable timekeeper, the sun was still required as a means for checking the accuracy of watches. In December 1843, James Bloxam patented the design for the optical parts of an instrument for determining the time at which the sun was due South. He sold the patent to Edward Dent, a London clockmaker, who made the 'dipleidoscope' and also published two booklets on the use of the instrument. While not always usable as a sundial, it occasionally carried a short hour scale for telling time between the hours of 9am and 3pm. However, its main purpose was for finding solar noon, which could then be converted into the mean time required by mechanical timepieces, by referring to a table of the equation of time.

The essential part of the dipleidoscope is a triangular 'prism' consisting of two mirrors and a piece of plane glass. This is set at an angle equal to the latitude and the instrument is positioned by means of a compass (either integral to the instrument or separate) so that the prism lies in the meridian plane. The two mirrors give two images of the sun, except when the sun is directly in the South, at which point the two images will coincide to form one. At this moment the time shown on a watch must be noted and checked against a table of the equation of time to find out whether it is accurate or not. The determination of noon by this method is accurate to the nearest second.

Those dipleidoscopes which have short hour arcs work by rotating the pillar that carries the prism around the hour arc (whose axis is provided by the pillar) until the images of the sun coincide, at which point a marker on the hour scale will indicate the correct time. A very few dipleidoscopes are also adjustable for latitude (like the one shown on page 122) and could therefore be used by travellers. All of the instruments which were intended to be portable rather than fixed in position were fitted with a compass, with a degree scale for setting the compass for the magnetic variation. Usually these were contained in rectangular boxes rather than circular bowls, since it was only important to show the few degrees to either side of North.

The dipleidoscope marks the changeover from sundials to watches as the latter became more accurate and more widely available. Thereafter sundials declined very rapidly as portable instruments, although they were still very popular as focal points in gardens and on the walls of churches. In the next chapter we will look at the fate of the portable sundial in the twentieth century.

Chapter Eight

THE SUN ECLIPSED

Magnetic dial, English, early twentieth century.

By the end of the nineteenth century, the watch had superseded the portable sundial as the most accurate personal time-teller. Its popularity grew while sundials lapsed into obscurity. The only area in which they seem to have held an interest as working time-keepers was as instruments for children's use. The magnetic dial which was discussed in the previous chapter appears to have become a common dial for such practices – it was easy to use and showed the principles of employing the sun as a timekeeper, though the small dials that were produced cannot have been at all accurate and so must have bolstered the superiority of the watch.

During the 1920s a company in America patented and manufactured a reasonably dependable and simple sundial, which was marketed as a 'Sunwatch', not as a sundial. The Outdoor Supply Company's dial seems to have been made specifically for children and came with a list of instructions for its use. Unlike the sundials of earlier years it was clearly a cheap instrument, being made of painted tin and carrying a very small compass of the sort which had become common for children's use. The dial was very simply made with three hour scales for latitudes of 35°, 40° and 45° N. The gnomon could be adjusted for these latitudes, being pivoted in the centre of the hour plate and marked for the three latitudes, but no others. However, an estimate could probably be made for any of the latitudes lying between these three numbers.

The most interesting aspect of this dial is the table which appears in the lid. This lists a large number of the cities in the United States; many of these are state capitals, but various other important places are included, such as Washington, DC and Los Angeles. As one might expect, the latitude appears, but only as the second attribute of each city followed by two further columns. The first information to be provided is the magnetic variation of each city, so that the compass of the dial can be set correctly. This is followed by both the latitude and the longitude – the latter being

'Sunwatch' made by the Ansonia Clock Company, 1920s.

essentially irrelevant to the procedure of finding the time. The column headed 'COR. MIN.' provides the allowance which must be added to or subtracted from the time measured by the sun in order to give the time according to the time zone within which the city lies. However, adjustment must also be made for the difference between solar time and mean time: this information is provided in a second table beside the latitude table, which lists the correction to be made for thirty-six dates through the year.

So this sundial provides all that is required for reading the solar time correctly and then translates it not just into local mean time, but into the mean time of the time zone in which the city is located. Time zones had been introduced in America during the 1870s, during the vast expansion of the railways. At that time it was realised that, with such a fast means of transport, it was not possible to use local time everywhere. Nor could the United States make use of one general time, because it covered such a wide area from East to West that a sensible mean time established for the East coast would be quite ridiculous for the cities of California. The end result was the division of America into five separate time zones, roughly corresponding to 15° of longitude each. This division of the continent is explained in the booklet which accompanies the Sunwatch and its relevance to the use of the instrument is also explained.

Such basic dials for children were essentially the last portable sundials to be manufactured for everyday use. However, the making of such sundials has not completely died out. The reason for this is the interest which has arisen in collecting old sundials and the appearance of sundials in museum collections relating to time measurement or to science in general. It is clear that some sundials have always been preserved for their elegance of design and their links to previous generations. This is the way

Lewis Evans, painted by W.E. Miller, early twentieth century.

Sir James Caird (1864–1954).

Robert Whipple, painted by Mary Marriot, 1930s.

in which many of the sundials of the fifteenth, sixteenth and seventeenth centuries have come down to us. However, towards the end of the nineteenth century an interest arose in collecting all forms of this now outdated time-teller. Perhaps it was because of the very 'quaintness' of the objects, once they no longer served a workaday purpose in people's lives. It was not long after the demise of the portable sundial as a common timekeeper that the first collections of sundials began to be brought together. One of the earliest and most important collections was that of Lewis Evans.

Lewis Evans came from a long line of people with interests in the mathematical sciences – one of his forebears had been an assistant at the Royal Observatory at the end of the eighteenth century. He also inherited his father's interests in archaeology and the family passion for collecting old artefacts. Why he chose to collect sundials is unclear, though he once remarked to an interviewer that he did so because the rest of his family collected all the normal things and that this was something unusual which did not clash with his relatives' interests. His earliest instruments were acquired when he was eleven. He was interested in many mathematical instruments, but his collection was largely limited to astrolabes and portable sundials. The instruments themselves were supplemented with an extensive collection of writings on mathematics and mathematical instruments, from the fifteenth century through to the nineteenth century.

Lewis Evans donated his collection to become the foundation of the Oxford Museum of the History of Science in 1924. As other museums relating to the history of science were established in England they too included large numbers of sundials. At Greenwich the National Maritime Museum was greatly indebted to the generosity of Sir James Caird, who presented a very large number of paintings and drawings relating to naval life, but also instruments such as globes, astrolabes, navigational instruments and sundials. Robert Whipple, whose collection was the basis for the Whipple Museum of the History of Science in Cambridge, amassed a store of early scientific instruments: while his main passion was for microscopes, his purchases included many other forms of instruments, including portable sundials.

This interest in collecting old sundials continued through the second half of the twentieth century, spurred by a rising interest in the history of science and also by museums devoted to the exploration of related areas. Many old instruments have come to light but, as with all such antiques, the increasing interest in sundials has led to the appearance of a considerable number of fake objects – sundials made in the twentieth century purporting to date from much earlier times. Sometimes these objects are easy to spot: they are poorly made with a lack of attention to accurate scale engraving and no understanding of the principles behind the sundials. At other times the difference between an original instrument and a forgery is harder to detect and numbers of these have been bought in good faith and have even made their way into museum collections. This compendium purporting to be by Christoph Schissler was bought by the National Maritime Museum as a genuine instrument, only to be identified as a fake as soon as it was compared with authentic works of the maker.

However, another form of imitation sundial has appeared in the form of replicas which are made both by commercial firms and by museums. Sometimes such objects are produced as promotional pieces for a company: an example is the ring dial shown here, thousands of which were made to advertise the products of the German firm Piz Buin. Yet large numbers are also produced to allow the public to have access to cheap versions of the original objects and to come to an understanding of their use. Many of these are not strictly replicas but facsimiles since they are not made of the original materials. Wood and card are common substitutes used for producing cheap copies of the original instruments, but it is not particularly rare to find them manufactured in brass as well.

Replica ring dial made as part of an advertising campaign for Piz Buin.

Twentieth century fake sundial, dated 1757. This is an obvious example of a forgery, since it includes nonsensical scales and is a very peculiar design.

Twentieth century fake sundial, inscribed Christophorus Schissler, 1565. This forgery was more difficult to spot, with high quality engraving and no obvious flaws in design. However, it is not like other instruments by Schissler and the engraving on the lid is taken directly from a popular 16th-Century book on instruments.

This cardboard replica of a nocturnal and horary quadrant (see p. 130) was sold by the Museum of the History of Science in Florence in 1974. It is a copy of a brass instrument made by the sixteenth-century Italian maker Girolamo della Volpaia. The front of the instrument carries a nocturnal which is used for finding the time at night by means of the stars. This side also provides a table for working out the length of time between sunset and midnight throughout the year. Such calculations allowed the time to be converted into 'Italian hours' – the system of time reckoning where the day was divided into twenty-four equal hours, beginning at sunset.

The back of the nocturnal carries an horary quadrant, which is used in much the same way as a Gunter quadrant, although the scale for setting the bead is placed down the side of the instrument, instead of along the circumference. An inscription gives the latitude of engraving as 43° 30′ N – close to the latitude of Florence and probably indicating the use of the instrument for that city. The replica instrument is accompanied by a booklet which describes the use of the nocturnal (though not the quadrant) and explains how adjustments can be made so that the nocturnal will provide accurate time now, despite calendar changes and alterations in the positions of the stars.

This type of cheap replica is becoming increasingly common and is important in providing people with an inexpensive means of owning sundials for themselves and understanding how these objects work. Other instruments which have been reproduced in similar ways include diptych dials, small horizontal dials (such as the replicas made from the dials found on the *Mary Rose*) and other nocturnals.

Facsimile nocturnal (left) from the Museum of the History of Science in Florence. The original instrument was made by Girolamo della Volpaia in 1568. A similar nocturnal (dated 1516) by another member of the della Volpaia family is shown on the right here. Both instruments are shown with the quadrant visible.

Our journey through the history of portable sundials has seen them pass from objects of little sophistication through their heyday in the seventeenth and eighteenth centuries and the lapse into obscurity during the nineteenth century to the returning interest in these timekeepers during the latter half of the twentieth century. Perhaps it is no longer true (as Emerson once wrote) that we 'should never be without a dial' but I hope that this book has shown the importance of portable sundials in the history of timekeeping and their worth as objects of beauty and interest.

GLOSSARY

altitude: the angular distance of an object above the horizon.

altitude dial: a sundial which uses the height of the sun in the sky to indicate the time.

altitude quadrant: a scale found on the back of some universal equinoctial ring dials. It is used to measure the altitude of the sun.

analemmatic dial: a sundial composed of a horizontal dial and an azimuth dial, which will show the correct time when both dials are indicating the same time.

armillary sphere: an instrument acting as a model of the heavens. It consists of a series of rings representing the major circles of the celestial sphere – equator, tropics, meridian, ecliptic, etc. centred around a ball or globe which represents the earth.

astronomical compendium: an instrument combining various different functions in one. It will almost always include a sundial of some form. Other functions may include a nocturnal, a latitude table, a lunar volvelle or a perpetual calendar.

Augsburg dial: a form of universal equinoctial dial produced in Augsburg during the seventeenth and eighteenth centuries.

azimuth: the angular distance of an object along the plane of the horizon from a fixed reference direction (usually North) .

azimuth dial: a sundial which uses the local direction of the sun (its azimuth) to tell the time.

Babylonian hours: 24-hour system of equal hours reckoned from sunrise.

Butterfield dial: a form of semi-universal horizontal dial, which is characterised by a bird-shaped pointer positioned against the latitude scale.

celestial equator: the equator of the celestial sphere; a projection on the celestial sphere of the Earth's equator.

chalice dial: a sundial in the form of a Mediaeval chalice. It may be in the form of an altitude dial or a direction dial.

co-latitude: the value obtained when the observer's latitude is subtracted from 90°. It is used in preparing the *quadrans vetus* for use.

compass dial: a form of horizontal dial where the dial plate is set directly over the compass.

compass rose: a circular pattern showing the principal directions of the compass.

cylinder dial: also known as the pillar dial. A form of altitude dial where the hour lines are inscribed on a wooden or metal cylinder

declination: the angular distance of a celestial object North or South of the celestial equator, measured along a great circle passing through the poles of the celestial sphere.

dipleidoscope: a noon-indicator invented by Edward John Dent in the mid-nineteenth century, and used for calibrating watches at midday.

diptych dial: a sundial formed of two leaves of wood, metal or ivory hinged together. It normally carries a horizontal dial on the inside of the lower leaf and a vertical dial on the inside of the upper leaf. Various other instruments may be included.

direction dial: a sundial which uses the direction of the sun with respect to the Earth's axis to tell the time.

dominical letter: the letter marking the first Sunday in a given calendar year. The dominical letters run from A to G where A is January 1. It is involved in the calculation of the date of Easter.

ecliptic: the great circle traced out by the sun in its yearly path through the stars.

epact: the moon's age at the beginning of the year. The value can be anything from one to twenty-nine. Epact tables are used for calculating the date of Easter and other moveable feasts in conjunction with the dominical letter and the golden number.

equation of time: the difference (measured in minutes and seconds) between the local apparent solar time and the local mean time; the earth does not move at a constant rate throughout the year and thus the length of time from one midday to the next as measured by the sun is not exactly twenty-four hours, but varies with the date. Thus solar midday can vary from the local mean midday by as much as 14.4 minutes in one direction and 16.4 minutes in the other.

equinoctial dial: a form of direction dial where the hour plate is set parallel to the equator, with the hour lines marked out at regular intervals around the plate.

equinox: the point in the year when day and night are equal in length (i.e. both of twelve hours' duration).

gnomon: a rod or pointer set on a dial, whose shadow shows the local apparent solar time.

golden number: a number marking the position of the current year in the Metonic cycle. This cycle is based on a lunar calendar within which the relative phases of the moon return to the same dates every nineteen years. The golden number is used in conjunction with the epact and the dominical letter for calculating the date of Easter. Also known as the prime.

Gregorian Calendar: the calendar introduced by Pope Gregory XIII in 1582, revising the Julian Calendar in order to keep the date in step with the vernal equinox.

horizontal dial: a form of direction dial in which the hour plate is set parallel to the horizon of the observer.

Gunter quadrant: a form of altitude dial consisting of a quadrant inscribed with hour lines, a pair of sights for viewing the sun, and a bead on a string to mark the time.

hour angle: the angle between the meridian line and the line engraved on a dial for a particular hour. The hour angle for a specific hour will vary depending on the latitude.

hour plate: the surface of a sundial on which the hour lines are inscribed.

inclining dial: a form of horizontal dial which can be adjusted for different latitudes by raising or lowering the hour plate and gnomon.

inhiraf: the qibla of a place, as measured from the South in that particular place.

Italian hours: twenty-four-hour system of equal hours reckoned from sunset.

jifa: a measurement given on Islamic dials to indicate the quarter of the horizon of a particular place where Mecca can be found.

Julian Calendar: the calendar instituted by Julius Caesar in 46 BC consisting of 365 days in each year and an extra day in every fourth year. These values were not completely accurate and resulted in the movement of the date with respect to the vernal equinox; the disparity eventually led to the reform of the calendar by Pope Gregory XIII to create the Gregorian Calendar.

latitude: the angular distance of an object North or South of the equator measured along a meridian.

latitude arc: the divided arc appearing on some equinoctial dials for setting the dial to the correct latitude for use.

latitude table: a list of towns or provinces with their corresponding latitudes.

longitude: the angular distance of an object East or West of a previously determined prime meridian at which longitude equals 0. It is measured along the equator.

lunar volvelle: a device which when set to the correct age of the moon will allow conversion from the time as shown by the moon to the solar time.

magnetic azimuth dial: a form of dial which uses the local direction of the sun and the movement of a magnetic needle to show the time.

magnetic dial: a form of horizontal dial where the hour plate is mounted on a magnetic bar, so that it is self-orienting.

magnetic variation: the angular distance between magnetic and geographic North.

meridian: a great circle through the geographic poles of the world (i.e. due North-South).

navicula de Venetiis: a Mediaeval form of altitude dial made in the shape of a ship.

nocturnal: an instrument used to find the time at night by means of the stars.

perpetual calendar: this can take one of two forms. Either it is a device for marrying dates with weekdays: this normally consists of a table of weekdays and months which allows the days of the week to be calculated for any month of the year if the dominical letter is known; the months are indicated by numbers and one often represents March, not January; an alternative form consists of a volvelle marked with the days of the week which moves over a circular scale of days. The second form of perpetual calendar provides details of such events as Saints' days, the dates of movable feasts and dominical numbers for a particular range of years; they can also be calculators for finding the times of sunrise and sunset or the length of the day or night. This latter form of perpetual calendar is sometimes known as a perpetual almanac.

polar dial: a form of direction dial where the hour plate is set in the plane of the Earth's axis.

prime: see golden number.

quadrans vetus: a form of universal altitude dial, which gives the time in unequal hours.

qibla: the direction to be faced in a particular place by a Muslim in order to be praying towards Mecca.

Regiomontanus dial: a form of altitude dial in which the time is told by a bead on a string. The bead is calibrated by means of an articulated arm moving on a grid of latitude and date lines.

right ascension: the angular distance of celestial objects measured eastwards along the celestial equator. It may be expressed in degrees or in hours and minutes.

ring dial: an altitude dial in the form of a ring, with a pin-hole through which the Sun can shine onto an hour scale on the inside of the ring.

scaphe dial: a dial in the form of a bowl.

shadow square: a device appearing on horary quadrants and other instruments for measuring the altitude of an object by means of ratios of lengths rather than angles; also known as the geometric square.

simple theodolite: a theodolite (angle-measuring surveying instrument) which measures only azimuth and not altitude. It consists of a degree circle around a compass, over which an alidade which can move around the circle is mounted.

solstice: the point at which the sun reaches its greatest value of declination during the year. The solstices occur in June and December, when the sun reaches the Tropics of Cancer and Capricorn respectively.

standard, common or astronomical hours: this hour system divides the day into twenty-four equal hours, in two groups of twelve hours, which began at midnight and midday. Further alternative names are German hours or French hours.

trigon: a grid used on the Regiomontanus dial for marking out the possible latitudes and dates of use.

unequal hours: the system of unequal hours divided daylight hours into twelve equal parts and the night also into twelve equal parts. Thus the length of the hours varied through the seasons as the days grew longer or shorter, and daylight hours were only the same length as night hours at the two equinoxes. Alternative names are temporal or canonical hours.

universal equinoctial dial: a form of direction dial which can be used in all latitudes and for which the hour plate is set to be parallel to the Earth's equator.

universal equinoctial ring dial: a form of direction dial consisting of two rings, one set parallel to the Earth's equator and one set parallel to the local meridian. The time is shown by a ray of sunlight falling through a hole in the central bridging bar onto the hour scale on the equinoctial ring.

vertical dial: a form of direction dial in which the hour plate is perpendicular to the horizon of the observer.

volvelle: a disc or discs which can be rotated over a fixed circular scale to aid in astronomical or mathematical calculations or conversions.

wind rose: a device consisting of the points of the compass arranged in a stylized star pattern and used in conjunction with a wind vane set at its centre to indicate the direction of the wind.

zenith distance: the angular distance of a celestial object from the zenith, the point on the celestial sphere which is directly overhead.

zodiac: sometimes used as an alternative name for the ecliptic, the zodiac is that band of constellations through which the sun appears to move in the course of a year. It is divided into twelve 'houses' each of which occupy 30° of the ecliptic and are symbolised by the constellation contained within that 30° stretch.

BIBLIOGRAPHY

Sundials have become a popular subject for books during the last twenty or thirty years. However, few of these books have extended sections on the history of the sundial – most are concerned with providing details of methods for constructing sundials, which has become a popular study. The following are some suggestions worth a look for people interested in making sundials:

René R.J. Rohr, *Sundials: History, Theory and Practice* (Toronto: University of Toronto Press, 1970, translated by Gabriel Godin). As well as providing instructions on making various sundials, Rohr gives one of the more detailed histories of sundials for the general reader, but he is not always accurate, and his information should be treated with care.

Albert E. Waugh, Sundials: Their Theory and Construction (New York: Dover, 1973): a clear exposition of sundial construction methods.

Some general information about the history of sundials can be gleaned from histories of scientific instruments, and four worth a look are:

Maurice Daumas, *Scientific Instruments of the Seventeenth and Eighteenth Centuries and their Makers* (London: Portman Books, 1989, translated and edited by Mary Holbrook);

A.J. Turner, *Mathematical Instruments in Antiquity and the Middle Ages. An Introduction* (London: Vade-Mecum Press, 1994);

Anthony Turner, *Early Scientific Instruments. Europe 1400-1800* (London: Sotheby's Publications, 1987);

G. L.E. Turner, *Nineteenth-Century Scientific Instruments* (London: Sotheby's Publications, 1983).

Catalogues of scientific instruments often include sundials and descriptions of their use and history. Two are mentioned below; two more worth a look are:

D.J. Bryden, *The Whipple Museum of the History of Science. Catalogue 6. Sundials and Related Instruments* (Cambridge: Whipple Museum of the History of Science, 1988);

F.A.B. Ward, *A Catalogue of European Scientific Instruments* (London: British Museum, 1981).

The British Sundial Society is a very active organisation and provides a bulletin three times a year, as well as meetings and an informative website at http:// www.sundialsoc.org.uk/

All sundials mentioned in the text which are in the collections of the National Maritime Museum are fully described in Hester Higton, *Sundials at Greenwich: a catalogue of the sundials, nocturnals and horary quadrants in the National Maritime Museum, Greenwich* (Oxford: OUP, forthcoming) [provisional title]

Chapter One

Information on early Egyptian portable sundials can be found in Sarah Symons, 'Shadow clocks and sloping sundials of the Egyptian New Kingdom and Late Period: usage, development and structure' in *Bulletin of the British Sundial Society*, no. 98.3

(October 1998), pp.30–36.

More detailed information on Roman portable sundials, and the Este dial in particular, is provided by Mario Arnaldi & Karlheinz Schaldach, 'A Roman Cylinder Dial: Witness to a Forgotten Tradition' in *Journal of the History of Astronomy*, xxviii (1997), pp.107–117.

There is one good general work on Greek and Roman sundials – Sharon Gibbs, *Greek and Roman Sundials* (New Haven: Yale University Press, 1976).

Recent discussion of the Canterbury Saxon dial can be found in Allan A. Mills, 'The Canterbury Pendant: A Saxon seasonal-hour altitude dial' in *Bulletin of the British Sundial Society*, no. 95.2 (June 1996), and the letters of Peter Drinkwater and Anthony Turner in subsequent issues.

Chapter Two

There is little information on either the *quadrans vetus* or the Regiomontanus dial. However, the *navicula* has sparked a great deal of interest and the following are just some of the articles available:

Christopher Daniel, 'The Sundial Page' in *Clocks*, October 1992, p.37;

A.W. Fuller, 'Universal Rectilinear Dials' in *The Mathematical Gazette*, XLI (1957), pp.9–24 (this also has a small section on Regiomontanus dials);

Ing. J. Kragten, *The Little Ship of Venice – Navicula de Venetiis* (Eindhoven: The Sundial Society of the Netherlands, 1989, translated by A.H. van der Wijck);

Kristen Lippincott, 'The Navicula Sundial' in *Bulletin of the Scientific Instrument Society*, no.35 (1992), p.22;

D.J. Price, 'The Little Ship of Venice – a Middle English Instrument Tract' in *Journal of the History of Medicine and Allied Sciences*, XV, 4 (October 1960), pp.399–407.

The chalice dial has been discussed briefly in Allan A. Mills, 'Chalice Dials' in *Bulletin of the British Sundial Society*, no. 95.3 (October 1995), pp.19–26.

Chapter Three

For more information on the development of the compass, see Colin A. Ronan, *The Shorter Science and Civilisation in China: an abridgement of Joseph Needham's original text*, vol. 3 (Cambridge: Cambridge University Press, 1995), chapter 1.

The history of the sinking and retrieval of the *Mary Rose* has been treated in various books. The best is probably Margaret Rule, *The Mary Rose: the excavation and raising of Henry VIII's flagship* (2nd edition, London: Conway Maritime Press, 1983) .

Chapter Four

The ivory sundials produced in Nuremberg are well documented. Two catalogues of major collections have useful historical introductory essays. They are:

Penelope Gouk, *The Ivory Sundials of Nuremberg 1500-1700* (Cambridge: Whipple Museum of the History of Science, 1988);

Steven A. Lloyd, *Ivory Diptych Sundials 1570-1750* (Cambridge, Massachussetts & London: Harvard University Press, 1992).

Information on the cities of Augsburg and London can be obtained from

Eugene F. Rice, Jr, with Anthony Grafton, *The Foundations of Early Modern Europe, 1460–1559* (2nd edition, New York & London: W. W. Norton & Company, 1994);

Mary Cathcart Borer, *The City of London. A History* (London: Constable, 1977);

Robert Gray, *A History of London* (London: Hutchinson, 1978).

The Humphrey Cole compendium was one of the subjects of an article by R.T. Gunther – 'The Great Astrolabe and other Scientific Instruments of Humphrey Cole' in *Archaeologia*, vol. 76 (1927), pp.273–317.

Chapter Five

For information on London during the seventeenth century, see the items listed under chapter four. Little information is available on Gresham College, apart from a recent brief anniversary publication – Richard Chartres and David Vermont, *A Brief History of Gresham College* (London; Gresham College, 1998). However, a biography of its founder is available: F.R. Salter, *Sir Thomas Gresham (1518-1579)* (London: Leonard Parsons, 1925).

Short biographies of Edmund Gunter, William Oughtred and Elias Allen can be found in the *Dictionary of National Biography* and have been extensively updated for the *New Dictionary of National Biography* (publication date, 2004). William Oughtred was also the subject of a biography by Florian Cajori – *William Oughtred: a great seventeenth century teacher of mathematics* (Chicago: Open Court Publishing, 1916). This has been supplemented more recently by information in chapter two of Frances Willmoth, *Sir Jonas Moore: practical mathematics and Restoration science* (Woodbridge: The Boydell Press, 1993). Elias Allen was the subject of the author's PhD thesis, which is in process of revision for future publication.

Chapter Six

The development and production of Dieppe magnetic azimuth dials is treated in Stephen Lloyd's *Ivory Diptych Sundials* (see listing for chapter four for full details). Information on the Huguenot wars in France can be found in various histories of France – see, in particular, Otto Zoff, *The Huguenots. Fighters for God and Human Freedom* (London: Allen & Unwin, 1943). A history of late seventeenth-century Paris is provided by Andrew Trout, *City on the Seine. Paris in the Time of Richelieu and Louis XIV* (Basingstoke: Macmillan, 1996), while details of Michael Butterfield's life can be found in Maurice Daumas, *Scientific Instruments of the Seventeenth and Eighteenth Centuries and their Makers* (London: Portman Books, 1989, translated and edited by Mary Holbrook). Information on the first Duke of Marlborough and his campaigns has been drawn from David Chandler, *Marlborough as Military Commander* (3rd edition, Tunbridge Wells: Spellmount Ltd., 1989) and Lt.-Col. G.W.L. Nicholson, *Marlborough and the War of the Spanish Succession* (Ottawa: Queen's Printer, 1955).

Chapter Seven

The list of works on the history of watches is extensive. Suggested items for further reading are:

Eric Bruton, *The History of Clocks and Watches* (London: Orbis Books, 1979);

Cecul Clutton and George Daniels, *Watches: A Complete History of the Technical and Decorative Development of the Watch* (London: Philip Wilson Publishers Ltd for Sotheby Parke Bernet Publications, 1979).

The invention and use of the dipleidoscope is discussed in Vaudrey Mercer, *The Life and Letters of Edward John Dent Chronometer Maker and some account of his successors* (London: Antiquarian Horological Society, monograph 13, 1977).

Chapter Eight

The collecting histories of Lewis Evans and Sir James Caird are described in A.V. Simcock, *Robert T. Gunther and the Old Ashmolean* (Oxford: Museum of the History of Science, 1985) and Kevin Littlewood and Beverley Butler, *Of Ships and Stars: Maritime Heritage and the Founding of the National Maritime Museum* (Greenwich/New Brunswick, N.J.; London: The Athlone Press, 1998), respectively.

INDEX